T0298526

Climate Change and Pragmatic Engineering Mitigation

Climate Change and Pragmatic Engineering Mitigation

edited by
Jacqueline A. Stagner
David S.-K. Ting

JENNY STANFORD
PUBLISHING

Published by

Jenny Stanford Publishing Pte. Ltd.
Level 34, Centennial Tower
3 Temasek Avenue
Singapore 039190

Email: editorial@jennystanford.com
Web: www.jennystanford.com

British Library Cataloguing-in-Publication Data
A catalogue record for this book is available from the British Library.

Climate Change and Pragmatic Engineering Mitigation

ISBN 978-981-4877-97-8 (Hardcover)
ISBN 978-1-003-25658-8 (eBook)

Dedication

To everyone who savors a tenable life.

Contents

4. Improving Heat Transfer Efficiency with Innovative Turbulence Generators **131**

Yang Yang, David S.-K. Ting, and Steve Ray

5. Effect of Ambient Temperature and Wind on Solar PV Efficiency in a Cold Arctic Climate **161**

Avinash Singh, Paul Henshaw, and David S.-K. Ting

Preface

To the worried minds, climate change is an ominous threat that is anticipated to imminently lead to the fulfillment of a doomsday scenario. Accordingly, some extremely high-risk, large-scale schemes, such as detonating massive explosions in the atmosphere to distribute dust to shield against part of the solar energy, and genetically altering the human species to become petite, less fertile, allergic to meat, etc., are arising. These radical ideas are driven by the fear that the many existing, and the new and emerging, balanced measures are too late and not forceful enough to mitigate the impacts of climate change. This volume intends to stimulate continuous scientific discussion across the diverse views, from the very conservative ones based on the notion that nature will take care of itself, to the radical, explosive ones. It is time for stakeholders, policymakers, and decision makers to be appropriately enlightened by the differing views. Let us try to forge some open-minded consensus to tread forward. Maybe the first step is to reduce the big problem into a lesser one, from which environmentally safe, low-risk remedies can be brought about to resolve it.

This volume commences with "Ocean, Weather and Climate Change" as Chapter 1. Former Commander of the Royal Navy, Graham T. Reader enlightens us with the connection between ocean, weather, and climate via the Great Ocean Conveyor. Reader makes it clear that a good understanding of the ocean is imperative toward a better appreciation of climate variability.

Human beings are bounded to dry land and everyone needs a roof over his or her head. In Chapter 2, "Environmental Impact Assessment of the Roof Insulation Materials during Life Cycle," Sua and Balo expound on the importance of roofs in terms of both economics and ecology. Four different roof insulation materials were analyzed based on eight characteristics, from global warming impact to water demand. It is soothing to learn that the most sustainable option, utilizing locally available sources can grant solid performance, along with the best cost and environmental values. Closely related to roofs is solar radiation, whose value is critical for smart and clean cities where solar energy is an integral part of the equation. Accurate estimation of available radiation is the key to capitalize solar energy fully. Balo and Sua studied two major cities in the Black Sea region to illustrate a comprehensive assessment of the suitability for profitable solar energy harnessing in Chapter 3, "Mathematical Radiation Models for Sustainable Innovation in Smart and Clean Cities." A solar energy conversion system is one of the many engineering innovations whose efficacy depends on the effectiveness of heat exchange. For solar photovoltaic panels, potent heat removal, for keeping the panels cool, is needed to overcome deterioration of their energy conversion efficiency with rising temperature. Yang, Ting, and Ray present "Improving Heat Transfer Efficiency with Innovative Turbulence Generators" in Chapter 4. It communicates the state-of-the-art of passive turbulence generators for augmenting forced convection. Designs that generate long-lasting thermal energy scooping streets of vortices appear most promising. Just when we thought there is not enough solar energy for applications in the far north, the frigid cold wind compensates this shortcoming by boosting the energy conversion efficiency by keeping the photovoltaic panels cooler than cool. Singh, Henshaw, and Ting illustrate this in Chapter 5, "Effect of Ambient Temperature and Wind on Solar PV Efficiency in a Cold Arctic Climate." What about for most of the population that resides in the warmer part of the earth? Wang, Henshaw, and Ting present a promising photovoltaic-thermoelectric hybrid power system in Chapter 6, "A Review of Current Development in Photovoltaic-Thermoelectric Hybrid Power Systems." The idea is to harness the thermal energy that negatively affects photovoltaic performance to produce high-quality electricity.

There is no doubt that it takes time to realize climate change mitigation. In the meantime, adaptation is essential. Permana and Petchsasithon propose a viable adaptation strategy in Chapter 7, "Low-Risk Engineering Adaptation Strategies to Climate Change Impacts at Individual Level in Urban Areas: A Developing Country's Viewpoint." The key is to call every individual to take part in "living with the flood," where every citizen is free to adopt one or more simple strategies. The simplest of which may be the

3Rs, Reduce, Reuse and Recycle. It is worth highlighting that the "living with the flood" approach has no potential to cause climate change itself. On the topic of adaptation, it is important to note that individuals respond to the thermal environment differently. In Chapter 8, "Analysis of Gender Differences in Thermal Sensations in Outdoor Thermal Comfort: A Field Survey in Northern India," Kumar and Sharma address outdoor thermal comfort. Among the non-meteorological factors that influence our senses in an outdoor space, gender stands out. Their study is based on a survey conducted in the semi-arid Haryana region in Northern India. What about edible buildings in the future? Afsar, Estévez, and Abdallah demystify this in Chapter 9, "Urbanization and Food in the Biodigital Age." With bio-manufacturing, buildings can be constructed with appetizing materials. Not that you want to gobble down your house; rather, the authors are proposing self-sufficiency via a sustainable design process loop. This volume wraps up with Chapter 10, "How Hydrogen Can Become a Low-Risk Solution for a Climate-Neutral Denmark by 2050?" by Wierciszewska and Xydis. This chapter specifically examines if hydrogen cars could influence Denmark in achieving their sustainability targets. As in our striving with climate change mitigation, its success leans not only on technological advancement but also on public perception and politics. Educating the public can affect both public perception and politics and, hence, the paramountcy of efforts such as the proper compilation of state-of-the-art vade mecums such as this volume.

David S.-K. Ting and Jacqueline A. Stagner
Turbulence & Energy Laboratory
University of Windsor, Canada

Acknowledgments

This volume was preordained via an opportune encounter with Jenny Rompas, an amiable publisher. The editors are indebted to the experts who penned the respective chapters that pieced together this timely treatise. A round of applause goes to the behind-the-scene reviewers who cross-examined the writing, and provided constructive comments that improved the quality of the manuscript. Providence from above carried this endeavor from inception to fruition.

Chapter 1

Ocean, Weather, and Climate Change

Graham T. Reader

University of Windsor, Ontario, Canada

greader@uwindsor.ca

To survive, humans, like other mammals, need to breathe air and drink palatable water. While some mammals live in water, humans inhabit the land, which covers about 29% of the Earth's surface, the remainder being almost entirely oceans, which contain undrinkable, salty, water. Not surprisingly then humans have been more focused on the habitability and resources of the land. Oceans have provided the means for transit and food, but little interest was shown in the physical, biological, or chemical structure of the oceans, until relatively recently. However, in the 15th and 16th century longer ocean voyages became more frequent and by the later 19th century undersea communication cable laying rapidly expanded. These activities demanded a far more detailed knowledge of ocean surface winds and currents, and the topography of the deep seas. Voyages of scientific discovery, and hydrographic surveying of the oceans, began to fill in the gaps. Ocean currents were mapped and eventually measurements started to be taken of deep-sea physical properties.

Climate Change and Pragmatic Engineering Mitigation
Edited by Jacqueline A. Stagner and David S.-K. Ting
Copyright © 2022 Jenny Stanford Publishing Pte. Ltd.
ISBN 978-981-4877-97-8 (Hardcover), 978-1-003-25658-8 (eBook)
www.jennystanford.com

Even so, linkages between the oceans and terrestrial weather were only firmly established three decades ago when it was recognised that ocean water circulations are interconnected by the *'Great Ocean Conveyor'*. This phenomenon carries thermal energy around the globe regulating the Earth's climate and weather. However, there are worries that anthropogenic climate change is degrading the oceans, disrupting the conveyor's positive influences. Subsequently, the United Nations, recognizing that there is a need to improve scientific knowledge of the oceans, have announced a decade of ocean science initiatives.

1.1 Introductory Remarks

This chapter is written primarily for a broad, tertiary level, readership seeking insights into ocean characteristics and the role they play in global weather and climate. Professional and scholarly oceanographers, ocean and Earth scientists, and their students, will already be familiar with much of the subject matter. The chapter begins with an abridged overview of how the existing understanding of the ocean's characteristics evolved from historic observations, and more recent measurements, presented in the context of the foundations of the scientific disciplines of oceanography and limnology[a]. The 20th century realization of the roles these characteristics play by means of the *'Great Ocean Conveyor'* interacting with the atmosphere's *'three cells'* to form global weather patterns and regional climates is explicated. The major ocean circulatory currents (gyres) and the most significant seasonal, reversing, regional winds (monsoons) are identified along with the observed impacts of regular, but non-seasonal, oceanic sea surface temperature anomalies such El Niño and La Niña and their interaction with atmospheric circulations [1]. The socioeconomic impact of all these phenomena on weather events and forecasting, especially regarding precipitation and droughts, is examined. Finally, how the oceans and cryosphere[b] may be affected by anthropogenic climate change (acc) is then considered. Subsequently, the concerns

---

[a]The scientific study on inland waters.

[b]The cryosphere are those parts of the Earth that are so cold that the waters are frozen solid.

associated with abrupt climate change, rising sea and temperature levels, and ocean acidification, are addressed [2].

It is highlighted that when attempting to describe the relationships between the ocean and atmospheric air circulations, and their influence on global weather and climate patterns, while numerical models provide some very helpful insights, it quickly becomes clear that there is considerable reliance on observational data and measurements. Moreover, in the absence of extensive efforts to obtain far more empirical data the prospects for verifying real-time predictive global weather tools appear remote. Fortunately, for the ocean aspects, in this era of efforts to develop a more sustainable future for the planet, the 2021 to 2030 decade has been proclaimed by the United Nations as the "Decade of Ocean Sciences for Sustainable Development" [3]. Indeed, one of the landmark Paris Agreement's Sustainable Development Goals—SDG 14—is specifically focused on issues related to oceans and seas [4]. One of the underlying themes of these UN initiatives is the critical need to improve our scientific knowledge of the oceans through enhanced data collection, especially in deeper waters. This is a weighty task since only *"5% of the ocean floor has been mapped and only 1% at high resolution and for 99% of habitable marine areas we lack basic biodiversity knowledge"* [5].

At the same time as the UN SDG 14 initiative was announced, an international consortium led by the Nippon Foundation launched the *Seabed 2030* project aimed at completely mapping the ocean floor by 2030 with a features mapping resolution of 100 × 100 m [6]—a very ambitious project which, with existing technologies, the Seabed 2030 team have estimated will take at least 1000 ship-years, e.g., 100 ships for 10 years [6]. Just how realistic this objective is can be gauged, perhaps, by the project's claim that the UN's 2017 estimate [5] of only 5% ocean floor mapping has already been increased to 15–17% by June 2020 [7]. With more topographical, physical, chemical, and biological data, it should be possible to more precisely predict oceanic conditions and better evaluate the impact of climate change on the ocean ecosystem [5].

Why the need for so much more data when sophisticated modelling techniques are available? The basic issue with all modelling is the uncertainty around whether all the pertinent factors have been included in the model, i.e., the model does not identify the 'missing' variables. Observations and data can lead to recognitions that additional characteristics need to be incorporated into a model. But are not the equations of oceanic motion and seawater thermodynamics well-known? The answer is yes, but they can only be simultaneously solved using numerical grid approaches and these require the use of approximations and parameterization techniques, which, in turn, require observed data [8]. For global weather forecasts and climate predictions, a similar set of 3-D equations describing atmospheric circulations over both land and sea also need to be solved. These are daunting tasks and it has been estimated that just for real-time ocean motion predictions the supercomputers in present use will need to become at least 10 billion times faster [8, 9]. Thus, although global weather forecasting has consistently improved in this century so that precise forecasts of about 10 days in advance are possible with numerical techniques from the time that quality initial data is available, there appears to be a time limit to this type of forecasting weather of about 2 weeks [10].

So, what have been the pathways to the accumulation of knowledge about the oceans?

1.1.1 Ocean Characteristics: Observations and Measurements

Physical evidence indicates that the oceans are at least 3.3 billion years old, but they may have been on the planet 'Earth' since the initial formation of our planet some 4.5 billion years ago [11]. The global oceans cover considerably more of the Earth's surface than the lands, they are on average deeper than the land is high, and they have a volume almost 10 times that of the land above 'sea-level'. Frequently the terms oceans and seas are used interchangeably to describe oceanic regions but the United States National Oceanic and Atmospheric Administration (NOAA) suggest that there is a difference in geographic terms between an ocean and a sea, in that seas are smaller than oceans and are partially enclosed by land [12]. In this chapter, both descriptions will be used. Moreover, although it is common to talk about the oceans as being regionally separate, i.e., the Pacific Ocean, Arctic

Ocean, Atlantic Ocean, Indian Ocean, and Southern Oceans, these are all interconnected in a single continuous ocean. However, while the somewhat contrived divisions are descriptively convenient when discussing ocean characteristics, for many reasons the waters contained in different regions do not mix well at their observed interfaces.

The increasing awareness of the relationships between the oceans, weather, and climate is a focus of this chapter. Why should we be interested in the oceans? Three obvious reasons are resources, trade, and global communications (undersea cables) routes and an important fourth reason is weather. We now know that weather on land is related to ocean characteristics and the interactions with atmospheric conditions can cause droughts and floods on land severely affecting agricultural activities and habitability for humans and animals. Humans are land-dwellers but historically many have always lived close to the ocean's coastlines. This is still the case today since according to the United States National Aeronautics and Space Administration (NASA) some 2.4 billion people live within 100 km of the world's approximately 620,000 km of oceanic coastlines [13]. Why? The oceans contain an abundance of food resources, and to unload these and other goods, harbours and port facilities have been necessary for several millennia. Today almost 90% of world trade is carried by shipping [14]. As these ports have developed, the number of people living in their vicinity has increased so that many of the modern, populous, cities worldwide are on or close to the oceanic coastlines. But archeologists have discovered numerous the remains of cities and communities under the seas which were overwhelmed by sea level rises likely caused by earthquakes and other natural phenomena [15, 16].

In addition to food and facilitating trade, the oceans, or more accurately the ocean floors, have been exploited for their resources of liquid and gaseous materials since the early 20th century. In this century, mining companies have started to focus on deep sea mining for metallic ores extraction. But to whom do these resources belong? As the economic value of oceanic resources continued to increase, especially in the last century, the ownership of these resources became a contentious issue between nations. These disputes have to a large extent been resolved by the United Nations Convention Law of the Sea

(UNCLOS) [17] that eventually came into force in 1994 [18] and which, as of 2019 [19], has been ratified by 168 parties[c], a notable exception being the USA. A key element of the convention is the definition of an Exclusive Economic Zone (EEZ), shown in Fig. 1.1, which defines an area 200 nautical miles from a country's coastline as being their sovereign territory. To implement and occasionally enforce such 'laws' or 'treaties', it is crucial to have accurate maps or charts of the ocean and its coastlines. The key developments in the history of charting the oceans were due to the 2nd century Greek philosopher Claudius Ptolemy's book *Geography*, or *Geographia* [20], and the 16th century Flemish cartographer and geographer Gerardus Mercator [21][d]. The so-called Mercator projection was developed to help ocean navigation and is still used today, and is more generally encountered as 3-D Globes.

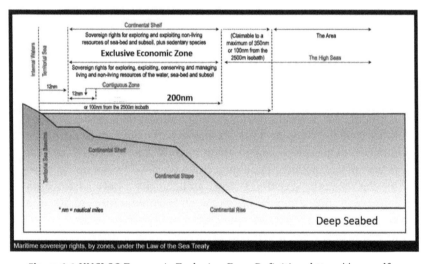

Figure 1.1 UNCLOS Economic Exclusion Zone Definition. https://www.dfo-mpo.gc.ca/science/hydrography-hydrographie/UNCLOS/index-eng.html.

Measuring the depth of seas and oceans using a line and a sinker mass was a technique familiar to early human civilizations [22]. Correlating these data with 'Mercator' based charts gave rise

---

[c]In UN parlance a party is a Country, State or other recognised entity such as The European Union.
[d]Born Geert de Kremer.

to science of hydrography in the 18th century. Knowing the shape and size of the oceans and proximity to land masses is crucial for navigation. Knowledge of the sea conditions, wind, waves, and swell[e], i.e., weather, likely to be encountered in an ocean voyage is crucial to the planning of a successful transit. Navigational charts do not provide this important information. The scientific study of the oceans, Oceanography, only dates to the last quarter of the 19th century CE[f], so how did sailors take cognisance of possible weather conditions? Mainly by observations and attempts to predict seasonal changes or at least catalog the observations from existing records. Human observations of tides were made in pre-history time and efforts to explain the phenomena of surface currents and the directions of prevailing winds encountered by mariners existed by the mid-16th century. Prior to that time, there were many earlier attempts to explain and forecast what today are called seasons and weather, dating back 5 millennia in India [23] and to the Greek Philosophers Thales, Hippocrates, and Aristotle [24] between 600 and 400 BCE. Aristotle published a four-volume treatise *Meteorologica* [25] which became the standard text for almost 2,000 years. Aristotle believed the 'weather' was due to the Earth being a sphere and the main influences were the four elements earth-water-wind-fire with the oceans and the atmosphere interacting in some way. He made some remarkable accurate observations about the reasons for 'weather' as well as significant errors such as wind not being the result of air motion [26]. Not surprisingly the focus of these early forecasting attempts was land-based weather.

In the last quarter of the past millennium, the era of the European Renaissance, scientists began to gain insights into the structure of the atmosphere using the innovative instrumentation developed to measure pressure, temperature, wind speed, and humidity [27]. These eventually proved to be the key parameters of 'weather'. It became clear that far more observations and data were required not only to develop hypotheses but also to test them. Scientists also suspected that the atmosphere somehow interacted with oceans and clouds. The oceans are largely opaque,

---

[e]Swell—"a long often massive and crestless wave or succession of waves often continuing beyond or after its cause (such as a gale)"; https://www.merriam-webster.com/dictionary/swell.
[f]CE = Common Era. Throughout this chapter, all dates will be taken to be either in the common era or before the common era (BCE)

and little was known about what existed below the surface of the seas, except that in 1843 the British[g] medic and biologist Edward Forbes calculated, based on observations of the top layers, that no marine life existed at depths below about 550 m (300 fathoms) [28]. This 'fact' was generally accepted for over two decades, but when scientists began to dredge the deeper seas, especially the Norwegian theologian and biologist Michael Sars and his colleagues, they found that marine life existed below 800 m [29]. The study of the deep, or more accurately the deeper seas, attracted an increasing number of scientists, but it was the successful laying of an undersea communications cable between England and France in 1850, that attracted the attention of governments and businessmen [30].

The British were interested in the concept of more rapid communications with all parts of its global empire while the Americans saw such cables as a means of expanding their trade and influence. In making such determinations the seminal event was the voyage of HMS Challenger, 1872–1876, a Royal Navy ship purposely modified for the scientific tasks of studying the physical, chemical, and biological characteristics of the oceans below the sea surface and into the deep seas [31]. The 50 volumes of reports produced from the global voyage provided the foundations for the modern scientific discipline of oceanography and its various sub-fields. Until this time regional oceanic climates had been recorded based on sparse observations, e.g., hot, warm, cold, windy, and so on, but the reasons for such climates were not known.

1.1.2   Winds and Currents

Although trade between global regions largely took place by means of roads and trails up to the 15th and 16th centuries, maritime trade was important in areas such as the Mediterranean as long ago as the 3rd millennium before the Common Era [32]. In the early 15th century the Ming dynasty Chinese Emperor Yongle built huge fleets of ocean going sail powered vessels and sent his Admiral, Zheng He, on seven major voyages reaching as far as the 'Persian' Gulf, East Africa, and Sri Lanka with the main aim to explore trading opportunities and spread Chinese

---

[g]Born in Douglas, Isle of Man.

influence [33]. These voyages were stopped after the death of Yongle, but three decades later the Fall of Constantinople to the Ottomans forced the Western European nations, especially the Portuguese and the Spanish, to find new routes by sea to China and the Spice Islands and it was during these 'Voyages of Discovery' that the Europeans 'discovered' the Americas [34]. As these voyages used sailing ships across wide expanses of ocean it became crucial to improve navigational techniques and identify wind patterns. The key patterns became known to sailors as the 'Trade Winds' which are prevailing easterly winds that circle the Earth close to the equator [35]. They would also discover what today is referred to as the Inter-Tropical Convergence Zone (ITCZ), a belt around the Earth approximately 5° of latitude north and south of the Equator where almost no wind would be encountered and became known as the 'doldrums' [36].

As whale oil became an increasing valuable commodity between the 16th and 19th centuries, especially for lighting, the whaling fleets of nations such as the United States of America and Britain increased in numbers and undertook longer voyages. Information of ocean winds and currents was recorded by the whalers adding to the information being collected by other sailing ships carrying cargo and mail between North America and Britain, and Britain and Europe. Ocean currents were identified as being as important as ocean winds and it was noted that voyages between different ports in England and the east coast ports of New York and Rhode Island, of the then American colonies, could take up to 2 weeks longer depending upon the port for embarkation and the destination port, although in some instances the longer voyages were over shorter distances [37, 38] The phenomenon the sailors were encountering was the 'Gulf Stream' which had first been observed and recorded by the Spanish explorer Juan Ponce de León in the early 16th century during a voyage to modern day Florida when, despite favourable wind conditions, an opposing surface current kept the ship almost stationary [39]. Spanish and other European voyagers did not readily share with others their knowledge of the Gulf Stream. An exception to this veil of secrecy was the chart and data published by the English and Colonial American ship captain William Hoxton in 1735 which was printed in London, UK, and although surprisingly accurate was largely ignored both then and now [40]. Hoxton used the term 'Northeast Current' on his chart and data publications but the description 'Gulf Stream' was used by colonial fisherman and eventually adopted by others in the latter part of 18th century.

Over three decades after the Hoxton chart, the polymath Benjamin Franklin, while the London, UK, based Postmaster General of British America, and who would become one of the founding fathers of the United States of America, received complaints about the significant differences in travel time of shipping crossing the North Atlantic. Franklin investigated the matter and, on his journeys between the English ports and British North America, collected data on wind, currents, and sea surface temperatures and in conjunction with a former Nantucket whaler captain Timothy Folger published a detailed chart of the Gulf Stream [41, 42]. The work of Franklin and Folger and others, such as the surveyor John De Brahm [43], allowed the Gulf Stream to be shown on maps, Fig. 1.2 [40], although it was only in the 19th century that the term 'Gulf Stream' regularly appeared on navigation charts.

Figure 1.2 Franklin's Gulf Stream Chart [40].

Having established its existence the next question was to be, *'what causes the Gulf Stream'*? Franklin suggested the stream was wind-driven and most others agreed. Franklin postulated that the direction of the stream would be the same as the prevailing wind and this remained the consensus for several decades. Franklin was partially correct in that ocean currents are produced by the winds blowing over the sea surface, but it was shown in the 19th century that the direction and speed of the currents do not match the overhead winds, largely through the published works of the American Naval Officer, Matthew Fontaine Maury which culminated in his seminal book *Physical Geography of the Sea* [44]. Based on his own observations during a 4-year global circumnavigation aboard the USS Vincennes and analyses of other observations taken from the logbooks of many whaling and transatlantic ships, he concluded that wind and current directions are not in harmony. Maury was also a strong advocate of shipping nations sharing their information on oceans and without doubt he was a pioneer oceanographer [45]. As the 19th century progressed, a standardized format for reporting wind directions and strengths became commonplace. The wind direction was described by the direction from which it blew, whereas water currents were described by the direction toward which they flowed—conventions which are still used today. While this may seem confusing, it is logical, as a wind coming from the south, for example, will drive the sea surface north and vice versa. However, it became clear from the growing collection of observations that the track of wind-driven surface currents was to the right of the wind direction in the Northern Hemisphere, but to the left in the Southern Hemisphere [46]. Why?

Several theories and hypotheses were proposed in the late 19th century to explain the relationship between prevailing winds and the accompanying water currents. These usually ignored earlier suggestions that the Earth's rotation played a significant role [47, 48]. This was to change thanks mainly to the efforts of Scandinavian scientists, especially the Norwegians Fridtjot Nansen and Vilhelm Bjerknes and the Swedish oceanographer Walfrid Ekman [49]. At the end of the 19th century, the legendary explorer and Nobel laureate Nansen provided oceanographic data from his many Arctic voyages to the physicist and fluid dynamicist Bjerknes and eventually to one of his pupils Walfrid Ekman [50]. The outcome was the seminal paper by Ekman which presented a mathematical model aimed at explaining Nansen's observations [49].

Ekman's 3D model of incompressible water motion included expressions to account for the friction (drag) between surface winds and subsequent water currents and a rotational inertial force identified by the French mathematician, engineer, and academic Gaspard Gustave Coriolis in the 1830s [51]. Ekman's elegant mathematical model provided an analytical explanation for the observations of Nansen, Maury, and others [44]. Although the analysis has continued to be developed since its first publication, the fundamental mathematical principles remain the same and the description '*Ekman Spiral*' has become part of the vocabulary of oceanographers and meteorologists, especially in university-level ocean sciences and oceanography courses. The spiral is illustrated on Fig. 1.3 [52] and although Ekman himself took some current measurements himself in 1931 it was only in the late 20th century that observations and statistical analyses confirmed the '*Ekman Dynamics*' [53].

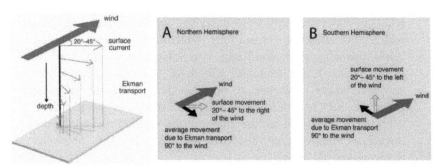

Figure 1.3 The Ekman Spiral [52].

While the work of the Scandinavian scientists enabled the relationship between wind and current direction to be established starting in the late 18th century and throughout the 19th century, sea surface temperatures (SST) and increasing deeper water temperature measurements were made by a number of ocean explorers and the general consensus was that temperature decreased with depth [54]. Moreover, along the track of an ocean current, SST temperatures were observed to change such as found by Franklin in his Gulf stream voyages. Did these temperatures play a role in oceanic current circulations was a question frequently considered by many 19th century polymaths and scientists of various persuasions, chemists, biologists, physicists, and geologists? Using, by today's standards, comparatively crude instrumentation, not only were temperatures measured, but also the 'saltiness' and density of the sea were recorded at the surface and at ever increasing depths [53, 55]. Despite the efforts of these early 'oceanographers' there remained a paucity of ocean wide observational data and the measurements that had been made were usually difficult to explain. For example, although not surprisingly, the SST measurements in equatorial regions were higher than in Arctic regions, the temperatures in the deep waters of both regions were very cold—a few degrees centigrade. The same phenomenon was observed in the deep regions at the western entrance to the Mediterranean. How could that be?

The 'new' sciences of thermodynamics and heat transfer could not be used to satisfactorily explain the vertical temperature differences as the deep-water temperatures were much lower than could be predicted. Maybe the temperature differences between the surface and the deeper waters were solely due to the depth of the water column? Although there was measured data to support this idea there were also indications that below a certain depth the temperature hardly changed, i.e., the temperature-depth gradient was significantly different at some point. Were there water layers with somehow colder water from more northern polar regions finding its way below the surface waters of the warmer equatorial areas? It was widely documented that surface currents, such as the Gulf Stream, carried warmer waters in the direction of the prevailing winds, but could there possibly be deeper water currents, as implied by Ekman's work, which carried colder waters in the opposite direction? Moreover, if there were deep water currents how were they caused? Winds drove the surface currents, but below the surface there were no winds and although Ekman's analysis explained how surface winds could affect currents a few hundred meters below the surface, what was driving the currents at the even greater ocean depths being found by 'lead and line' measurements especially by ships of the British Royal Navy [56]? These were just a few of the questions about the nature of the ocean currents that scientists were attempting to answer as the 19th century progressed, usually with insufficient data.

The increasing knowledge of ocean structures, water properties, and circulation patterns would eventually lead, by the end of the 20th century, to a better understanding of the roles played by the oceans in global weather patterns and climate both on land and at sea. Even so, at the start of the last century only hydrography—originally a systematic attempt to chart (map) the topologies of oceans—attracted the attention of governments and their navies, largely to aid ocean navigation and safe transits [57, 58]. Little government interest—reflected by levels of funding—was taken in the other ocean sciences and it was left to private sponsors and non-government bodies to help establish non-military maritime and oceanographic research institutions and societies such as the Scripps Institute of Oceanography [59], now part of the University of California at San Diego, and the Marine Biological Association laboratory in Plymouth, UK [60] in the late 19th and early 20th centuries. This was to change with the arrival of submarine warfare during the First World War and the consequent requirement to develop a method of detection when the submarine was submerged. By the end of the war acoustic (sound) detection methods, initially known as ASDIC but later as SONAR[h], had been developed [61].

However, almost a century before their use as a submarine detection method, sound reflections had been first used to measure the depths and underwater topology of Lake Geneva in Switzerland [56]. This type of lake depth measurement was repeated throughout Europe and in some States of America by scientists who became known as 'limnologists' and who studied the physical, chemical, and biological characteristics of lakes and other inland waters [62]. Throughout the 19th century, similar measurements to those being collected from the oceans were also taken for lakes of varying depths. By mid-century it was noted that, as with the oceans, there were water layers of different temperatures notably below a surface layer [e.g., 63, 64]. It was found that layers below the surface had measured temperatures where the decrease with depth was greater than the surface layer above and greater than the layer below. This notable temperature rate change was given the German name *Sprungschicht* but renamed 'thermocline' in 1897 by the American Limnologist E. A. Birge [65] and which would soon be adopted by the Ocean Sciences.

---

[h]SONAR – **SO**und **NA**vigation and **R**anging.

After the First World War, the government level interest, i.e., funding levels, in the ocean sciences were reduced, but scientific work continued helped by support from other financial sponsors especially at the already established centres. By 1930 the now most frequently and regularly citied research centre, the Woods Hole Oceanographic Institution (WHOI) located in Massachusetts, USA was founded [66]. As the Second World War approached, driven by naval requirements [67, 68], government interest again ramped up and today over 50 countries now have at least one oceanographic research center and usually many more, some are associated directly with universities and governments, others operate independently and at least one is a registered charity [69]. Many international ocean science partnerships have emerged since the latter half of the 20th century and recently the Seabed 2030 project [6] has become a touchstone example of such collaborative efforts.

Millenia of recorded seafarers' observations and almost three centuries of increasingly 'scientific' observations and data collection, since the detection of the Gulf Stream, enabled the Ocean Science disciplines to be established. These, often amazing, historical efforts coupled with the advent of 19th century instrumentation and more recently high-resolution sonar, satellite, and computer technologies, have provided the wherewithal for the modern understanding of how oceans work and their characteristics [9]. The awareness of the roles the oceans play in Earth's weather and climate systems in conjunction with atmospheric circulation is continually increasing although, as manifest in the UNESCO [3], UNFCC [5] and the associated Seabed 2030 [6] initiatives, the pace of this awareness needs to be significantly accelerated. But what knowledge has been assembled to date especially from physical observations and increasingly complex numerical models?

Before embarking on an answer to the previous question in the following sections of the chapter, it should be noted that a marvellous textbook, now in its sixth edition, together with an accompanying website of updated and supplementary materials, is an essential read for tertiary level students, faculty, and researchers interested in detailed studies of *descriptive* physical oceanography [70]. This book contains several hundred pages, many presentations, graphs, and pictures so it can be appreciated that by comparison this chapter provides only a snapshot answer to the question of accumulated knowledge.

1.2 Ocean Clines, Gyres, and Atmospheric Circulations

1.2.1 Thermocline, Halocline, and Pycnocline

By the start of the 20th century, recognition of the 'thermoclines' in lakes and oceans, and an increasing database of measurements, allowed an appreciation of the vertical thermal stratification of bodies of water to be gained. In simple terms, the vertical stratification in the open oceans was found to be typified by three depth layers[i], an upper zone[j] with temperatures like those of the surface, a zone below where the ocean temperature decreased rapidly, and a deep zone where temperature changes were much slower [70]. A typical ocean water temperature profile is shown in Fig. 1.4 [71]. The thickness (depth range) of the defined layers depends on geographical location as does sea surface temperatures (SST), which are largely driven by the amount and intensity of the solar radiation that a region receives, so not too surprisingly the oceanic equatorial regions have higher SSTs.

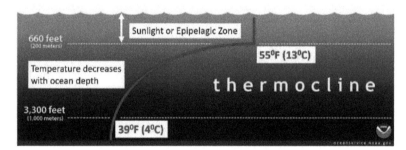

Figure 1.4 A typical ocean temperature profile [71].

---

[i]For several reasons ocean scientists today also further divide the deep layer giving a total of 4 and sometimes 5 zones.

[j]Usually referred to as the 'sunlight zone' or mixed upper layer but more scientifically as the *epipelagic zone.*

Generally, the higher the latitude the colder (lower) the SST, but there are exceptions, and this is the result of ocean currents redistributing heat energy. SSTs also change on a 24 h basis (diurnal) like the atmosphere above the surface, but the range is smaller due to liquids having a higher specific heat capacity than gases. The SST values are also seasonal because different parts of the Earth receive the Sun's most direct rays throughout the year[k] as the Earth, rotating on its tilted axis, orbits the Sun. Global measurement of SSTs are made daily and published almost in real-time. The data for 23 June 2020 is shown in Fig. 1.5 [72]. Such data is very useful for weather forecasters, particularly for indications that a known seasonal or non-seasonal feature is about to form, e.g., a hurricane or the onset of an El Niño 'episode[l]'.

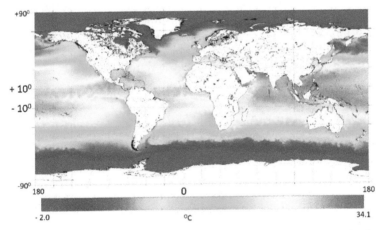

Figure 1.5 Global ocean surface temperature profile for 23 June 2020 [72].

There are other ocean clines[m], but the two important ones, in addition to the thermocline, are the halocline and the pycnocline. The first is concerned with the amount of 'salt' in the oceans—the salinity—which like temperature varies depending upon geographical location and is generally lower at the equator and polar regions, but higher at mid-latitudes. The 'saltiness' of

[k]365.26 days (1 Sidereal year).
[l]Official description when the phenomenon lasts for more than 7–9 months.
[m]Oceanographers use 'cline' to describe a fluid property gradient, but there are variants of the meaning among other scientific disciplines.

seawater is caused by dissolved 'salts'—ionic constituents—such as Chloride and Sodium which together account for just over 85% (by mass) of the ocean salts. On average there are 34.482 g of dissolved salts in each kilogram of seawater and of the numerous chemical elements and compounds found in ocean water there are six major constituents as shown in Table 1.1 [73].

Moreover, although the salinity varies across the oceans, in 1865, Dutch geologist Johan G. Forchhammer, after chemically analyzing samples from different oceans, hypothesized that the constituents are always present in the same relative proportions, but it was a more detailed analysis of all the major constituents by the Anglo-German scientist William Dittmar, using samples collected during the HMS Challenger voyages, that confirmed Forchhammer's theory [74]. Dittmar's or Forchhammer's principle is frequently referred to as the *Principle of Constant Proportions* [75] or the Principle of the Constancy of Composition [73]. Forchhammer is also credited with coining the term *salinity* [75].

Table 1.1 Major Ionic Constituents of Ocean Water.

Ion	% By mass of 34.482g
Chloride, Cl^-	55.04
Sodium, Na^+	30.61
Sulfate, SO_4^{2-}	7.68
Magnesium, Mg^{2+}	3.69
Calcium, Ca^{2+}	1.16
Potassium, K^+	1.10

Salinity is difficult, but not impossible, to measure directly and usually the electrical conductivity of the seawater is measured and a mathematical expression relating the two is used to determine the salinity, such as [76]; Salinity = 1.80655 × Chlorinity.

Where does the 'salt' come from? The ancient Greek philosophers [77] observed that seawater was 'bitter' in comparison with freshwater and suggested various reasons to explain the difference. These included 'salt-fountains' at the bottom of the sea [70], but Aristotle's contention was that the saltiness and bitterness of seawater was caused by "*an admixture of water and various earthy residues borne by rivers"* [78]. Aristotle's opinions remained unchallenged for almost two millennia, but as chemical analysis began to emerge from the 17th century onwards, coupled with the increasing collection of samples, a better understanding of the causes of ocean salinity was established. The river attribution of salts-in-the-sea proved to be correct, although not in the manner Aristotle proposed. The erosion of land rocks was identified as the source of the salts which were carried to the oceans by rivers. This erosion was boosted by precipitation as 'run-off' and because rain is slightly acidic [79]. However, in the 1970s, with the discovery of sea-floor hydrothermal vents it soon became clear that the ocean itself was also a source of saltiness, so the 'salt fountains' suggestion was proven to be not as farfetched as many, including Aristotle, argued. Nevertheless, the primary source of ocean salts is terrestrial rock erosion although chemical seepages from industrial, agricultural, and wastewater can distort local, regional, and, eventually, global measurements [80]. The latter, which may be labelled anthropogenic influences, are relatively recent as it is believed that the *"concentrations of elements in seawater … have remained virtually unchanged for more than 500 million years"* [81].

For many years the units used for salinity measurements were parts per thousand (ppt) and given the symbol ‰ but since the adoption of the Practical Salinity Scale (PSS) in 1980 the modern measurements are described in psu (practical salinity units) intended to reflect the method of measurement [82]. Much of the considerable literature on ocean sciences has used the ppt system, but the numerical values of the two systems are practically identical. For researchers and educators an interesting web discussion on the two systems is available [83]. Whatever the salinity measurement system, the overall concentration of the ion constituents varies with temperature, precipitation, and evaporation. Moreover, in a similar manner as the vertical temperature profiles the depth-salinity characteristics exhibit rapid value changes—haloclines—between the ocean layers above those of deeper waters, as illustrated in Fig. 1.6 showing vertical temperature, salinity, and density profiles [73].

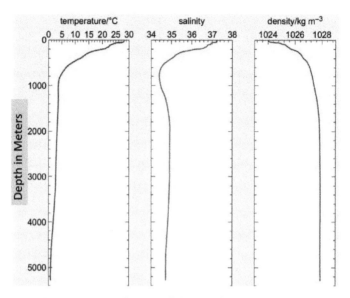

Figure 1.6 Temperature, salinity and density depth profiles at 20°S in the Atlantic Ocean [73].

The layer where the density gradient is at its greatest is known as the pycnocline. As 'Salty' water is denser than freshwater, and warm water is less dense than cold water, rapid changes in density may be due to decreasing temperatures or increasing salinity, if, for the former reason, the pycnocline is also a thermocline, or, if the latter, then the pycnocline is a halocline [84]. The pycnocline is a very stable ocean layer and as such generally impedes any turbulent mixing between the surface layer and the deep layer [70]. In turn this slows down deep-water movements, but near ocean bottoms, especially in colder regions, deep water currents exist which play important roles in both global weather patterns and climates. In the three-layer depth structure of the oceans the pycnocline and the surface mixed layer only exist between approximately 55° North and 55° South and are absent at higher latitudes. A few degrees above these limits, the officially defined polar regions start, as shown on Fig. 1.7 [84]. Seawater densities reach their highest values within these colder polar regions.

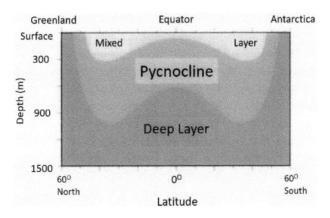

Figure 1.7 Global sea density layers [84].

Regardless of whether a pycnocline exists, the density of seawater at any location can be calculated using a universally agreed, but quite complex, 'Equation of State for Seawater', albeit simplified mathematical closed-forms are often used in modelling studies [85, 86]. However, in terms of pure physical relationships, seawater density depends upon the in-situ salinity and temperature. The difference in density between locations promotes relatively slow moving[n] ocean currents, which extend to and from the ocean floor, leading eventually to global patterns of 'thermohaline' circulations [87]. The main driving force of this type of circulation is the consequence of higher density seawater sinking into the deep waters of regions above north and south latitudes of 55° and greater. A schematic representation of the ocean thermohaline circulations is given in Fig. 1.8 [88]. These circulations are instrumental in carrying colder polar water towards the warmer equatorial waters, and warmer waters to the colder regions. They are also involved in the phenomenon of 'upwelling' which occurs when winds drive surface waters away from a region or a coastline and are replaced by deeper colder water. As these upwelling deeper waters are invariably rich in nutrients, where they occur usually coincides with high fish populations. The thermocline currents contribute

---

[n]In comparison to wind driven surface currents.

to general ocean water circulation, but their proportion is usually much lower than wind-driven surface currents. For example, the driving forces of the Gulf Stream, which has already been encountered as a large, persistent, and permanent current, are about 80% wind and 20% thermohaline effects [88].

Arguably, the Gulf Stream is the most celebrated of all observed ocean currents, but since the early observations of this phenomenon many other durable currents have been identified. In specific combinations, these currents, depending upon their location, can form large water circulation patterns known as Gyres as discussed in the next section.

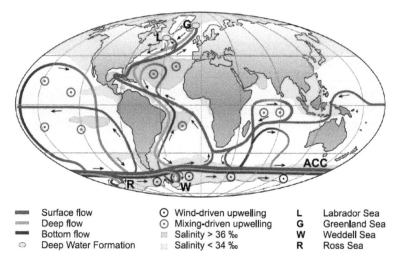

▬	Surface flow	☉	Wind-driven upwelling	**L**	Labrador Sea
═	Deep flow	☉	Mixing-driven upwelling	**G**	Greenland Sea
▬	Bottom flow	▦	Salinity > 36 ‰	**W**	Weddell Sea
○	Deep Water Formation	▦	Salinity < 34 ‰	**R**	Ross Sea

Figure 1.8 Ocean Thermocline circulation. *Note: ACC—Antarctic Circumpolar Current* [88].

1.2.2 Gyres

Gyre, literally a ring or circle°, is a term applied to large rotating ocean currents which move clockwise in the Northern Hemisphere and counter-clockwise in the Southern Hemisphere. Gyres are created by wind driven ocean currents, the Earth's rotation (Coriolis and Eckman effects), and the relative proximity of land masses [89]. The physical characteristics associated

---

°Latin, *gyrus*; Greek, *gyros*; a ring or a circle.

with a particular gyre can also be influenced by the 'clines', as the two are not wholly independent, and the overlying atmospheric pressure. Gyres are classified into three categories: Tropical, Sub-Tropical, and Subpolar. There are five major ocean gyres; the North Atlantic, the North Pacific, the South Atlantic, the South Pacific, and the Indian Ocean, as shown in Fig. 1.9 [90]. Wind-driven circulations dominate the regions where ocean gyres occur, especially in the surface layers. However, the circulations vary in speed, width, and depth. The faster surface currents, such as encountered in the northern Atlantic's Gulf Stream and the northern Pacific's Kuroshio Current, tend to be narrow bands of circulating waters, 50 to 75 km wide, akin to 'rivers-in-the sea' while the much slower surface currents such as the California Current can be literally hundreds of kilometers wide [91].

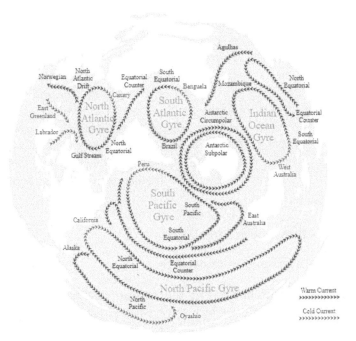

Figure 1.9 Major ocean gyres [90].

Gyres and thermohaline circulations constantly move large quantities of ocean waters carrying immense amounts of heat energy over considerable distances. The primary source of the heat energy input to the oceans (as well as land) is direct solar radiation which is a maximum in the equatorial regions resulting in higher sea surface temperatures as mentioned in Section 1.2.1. The minimum temperatures occur in the polar regions and thus, because of the physics of thermofluids and heat transfer, the warmer waters of the equator will move towards the colder polar regions and in doing so help regulate the Earth's climates. So why does colder water from the high latitude colder regions move towards the equator? Is it solely because of the direction of some of the wind-driven and density currents, i.e., gyres and thermohaline circulation? The answer is no, as atmospheric circulation patterns, e.g., the 'trade-winds' are also involved. But the connection between the two circulations is far from simple. For instance, while it is well-established that ocean surface wind directions and strengths are largely dictated by the prevailing atmospheric air characteristics it is also the case that ocean SSTs and mixed layer depths, in turn, influence atmospheric winds. However, this coupling is one of many between ocean and atmosphere and as stated in the UN's 'World Ocean Assessment report' [92], *"From the physical point of view, the interaction between these two turbulent fluids, the ocean and the atmosphere, is a complex, highly nonlinear process, fundamental to the motions of both"*. To try to comprehend how the ocean is involved in weather, climate, and climate change it is then also important to gain an appreciation of the fundamentals of atmospheric circulations.

1.2.3 Atmospheric Circulations

1.2.3.1 Layers of the atmosphere

The Earth's atmosphere, like the oceans, can be divided into vertical layers each of which have largely unique physical and chemical properties. These layers were discovered over the past 150 years or so by observation and measurement as instrumentation became increasingly sophisticated and accurate. Today, these layers, starting from the Earth's surface, are labelled in turn; the troposphere, the stratosphere, the mesosphere, the thermosphere, the exosphere, and the magnetosphere [e.g., 93]. To these can be added the ionosphere, which straddles several layers and contains strong concentrations of electrically charged particles which play a major role in global radio communications and where, as it overlaps with the thermosphere, the remarkable light shows known as the 'Northern Lights' (Aurora Borealis) and 'Southern Lights' (Aurora Australis) occur. All these layers have various thicknesses depending on the geographical location and the climatic season. The greater the distance from the Earth's surface the thinner—less dense—the atmosphere becomes, and at about 84 km above the surface it becomes too thin to support aeronautical flight because of the lack of generated lift. This occurrence is known as the Kármán Line in honor of Theodore von Kármán who calculated its existence [94]. There is no universally agreed height of this line and different jurisdictions place it from 80 km to 100 km. Why is this important? The line is often referred to as the edge-of-space and has legal ramifications since airspace below the line is deemed to be national airspace and any space above is reckoned to be free space. Satellites, spacecraft, and the International Space Station all operate above the Kármán Line, even the so-called low-orbit weather satellites [95]. Most commercial aircraft operate in the upper parts of the troposphere at altitudes around 12 km except for very short-haul flights.

Measuring atmospheric phenomena occurring between altitudes of 80 and 120 km, the top layers of the mesosphere and lower layers of the thermosphere, is very difficult, indeed some scientists lightheartedly call this 'the ignorosphere' [96]. However, when discussing the relationship between ocean and atmosphere and the influence of their interactions on weather and climate, all the layers or 'spheres' except the one closest to the Earth's surface—the troposphere—and parts of the contiguous layer, the stratosphere, can also be ignored, not in an insouciant manner, but because almost all weather occurs in the troposphere. Why should the stratosphere, the layer above the troposphere, and which reaches a maximum elevation of approximately 50 km, be of interest? The Bête Noire of Greenhouse Gases, Carbon Dioxide, is found in trace concentrations in the stratosphere [97], but it is its 'ozone layer', or sometimes its absence, that is the primary concern.

This stratospheric layer protects the lower atmosphere from the harmful effects of ultra-violet (UV) radiation, specifically a portion of the UV in a wavelength band known as UVB. If the effectiveness of this protective layer is diminished then more UVB reaches the troposphere, and eventually the Earth's surface, negatively impacting the health and well-being of humans, land and sea animals and ecosystems, and increased levels of photochemical 'smog' [98]. However, not all the 'smog' that is encountered is due to stratospheric ozone depletion. Tropospheric 'urban-smog' can be generated by the chemical interaction of air pollutant emissions, both natural and anthropogenic, with solar radiation. Smog events caused in a such manner may last for only a few hours or linger for several weeks, but have damaging effects on human health and agricultural production [99]. Thus, stratospheric ozone is beneficial to the Earth's ecosystem, tropospheric ozone is not.

The layers of the atmosphere and the associated atmospheric circulations and characteristics are extremely complex, probably more so than their ocean counterparts, and they are not yet fully understood. Fortuitously for any discussion of ocean-atmosphere weather and climate relationships, only the circulations in the troposphere need to be considered. Why? The troposphere layer contains 99% of global atmospheric water and about 75% to 80% of the atmosphere's mass. The weather experienced on the ocean and land is generated within this layer, whose average thickness is 10 km, but can vary between 7 and 18 km. Air circulation within the troposphere includes horizontal aspects, as manifest in the surface trade winds which flow towards the equator, and vertical characteristics due to temperature gradients. Explanations for why this was the case were proposed in the 17th century, perhaps the most notable being that by Edmund Halley, an astronomer of 'Comet fame' who suggested that the cause was the thermal effects attributable to the upward motion of solar-heated equatorial air, with colder surface air rushing in to replace the ascending air [100]. This explanation, while not wholly inaccurate, persisted until challenged by another Englishman, George Hadley who, unlike

Halley, did not dismiss rotational effects in his explanation and further proposed that in both hemispheres there was also only one form of air circulation.

Hadley posited that the warm air at the equator would rise and then flow towards the poles at high altitudes, but as the air approached the colder regions it would become denser, start to descend, and then flow back at lower elevations to the equator where the process would be repeated [101]. Hadley's hypothesis was a better explanation of the physical causes of the trade winds than Halley's. However, by the end of the 19th century it was determined, as more detailed knowledge was accumulated about the winds and vertical air properties in the higher latitudes, that Hadley's proposals did not adequately describe the air circulations encountered outside of the trade wind regions.

In conjunction with his attempts to explain the reasons for the trade winds, Halley also produced a wind-map of the trade winds between 30° North and 30° South of the equator, noting that outside this range the winds were variable [100]. But just how variable? As early as the 16th century it had been observed that prevailing surface wind directions changed outside the region of the trade winds from easterly to westerly, but within a zone at approximately 5° latitude either side of the equator the trade winds became much slower and sometimes non-existent—the doldrums [36]. The changing wind directions experienced by sailors outside the so-called *horse latitudes* of ± 30° also proved to be much stronger winds and were given revealing names by those that experienced them such as, the *'Roaring Forties'*, the *'Furious Fifties'*, and the *'Screaming Sixties'* [102] and these could not be accounted for using Hadley's notions. In the mid-19th century an American meteorologist, William Ferrel, using a mathematically correct Coriolis expression, was able to not only clarify the influence of the Earth's rotation on wind deflections, but also provide insight into the reasons for wind characteristics in the higher north and south latitudes above those of the trade winds [103].

As the 20th century progressed, several scientists started using kites and balloons to augment land based physical measurements of air properties at increasing altitudes. The immediate atmospheric layer above sea level was discovered and given the name the

'Troposphere' as well as the next contiguous layer called the 'Stratosphere'. Both these terms being given by the notable French meteorologist L. P. Teisserenc De Bort [104]. The boundary between the two layers was named the 'Tropopause' by the British railway engineer turned meteorologist W. H. Dines [104, 105]. The scientific investigations of Hadley, Ferrel, and De Bort established the existence of global atmospheric circulation cells, the effects of which had been literally described and observed by sailors since at least the 15th century. So, what are these cells and how is the tropopause involved?

1.2.3.2 The Three cells and the tropopause

As a per unit area, the equatorial regions receive more solar radiation than at the two polar regions; air and sea surface temperatures are at their warmest, whereas they are their coolest in the higher latitudes. The Second Law of Thermodynamics stipulates that heat flows from a high temperature to a low temperature, so does this mean that the surface air at the equator flows directly over the ocean and land to the poles? The answer is no, not directly. Why? In the troposphere the air pressure and temperature are at their highest at sea level, then both decreases with altitude. The air temperature reaches a minimum at the interface layer of the troposphere and the stratosphere—the tropopause—whose thickness and elevation is far higher at the equator than the poles [106]. The air heated at the equator rises vertically, but once within the tropopause it becomes less buoyant and, driven by winds, it starts to flow towards the poles. However, during the flow the air temperature drops, and it becomes denser and sinks back to the surface, not at the poles, but at the northern and southern latitudes of about 30° before eventually returning to the equator. In this way, the warmed air from the equator is prevented from directly reaching the poles [107]. This wind driven air circulation between the equator and the lower latitudes is named the Hadley Cell, in honor of George Hadley.

At the poles cold dense air descends and flows at low altitude in the general direction of the equator but in doing so the air temperature starts to rise and at latitudes of about 60° in both hemispheres the air ascends and flows back towards the poles.

These are the polar cells [81]. In the region between the edge of the Hadley cell trade winds at approximately 30° latitude, north and south, and the start of the polar cells in the vicinity of 60° latitude, north and south, another circulation cell exists which not surprisingly is referred to as the mid-latitude cell, although the term Ferrel Cell is popular [108]. The three cells' structure along with the ITCZ is shown in schematic form in Fig. 1.10 [109]. Although these cells are features of the troposphere their heights are dictated by the tropopause, which acts as a decreasing ceiling between the equator and the poles, as shown in Fig. 1.11 [110].

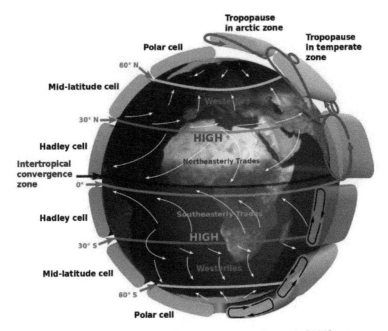

Figure 1.10 The three-cell atmospheric structure schematic [109].

These atmospheric artefacts largely dictate the global climates over land regions by not only transporting heat energy from the equator to the poles, but also by creating semi-permanent regions of high and low pressures. When heated air ascends, it creates low-pressure areas, whereas cold descending air creates regions of high pressure. Low-pressure areas especially around the equator are associated with high levels of rainfall and cloud formation. The high-pressure areas formed at the edges of the north and south Hadley cells and in the polar regions generally experience far less precipitation and cloud cover. The circulating cells then effect both short-term weather and longer-term climates. The latter are classified using the Köppen–Geiger (K–G) climate system, which has five major zones ranging from tropical to polar and numerous sub-zones within each of major regions. These zones and sub-zones are largely based on the types of vegetation that can be grown within them [111]. Based on vegetation-based empirical evidence collected and collated by Köppen at the end of the 19th century, the system has been revised several times since then and, despite its limitations, remains the most popular system [112, 113].

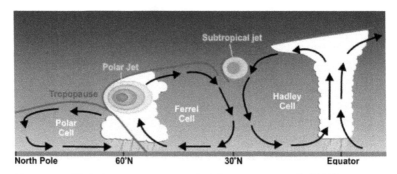

Figure 1.11 Variations in tropopause height equator to pole [110].

Care must be taken however when the system is used to describe ocean climate characteristics. For example, up to the latter part of the 20th century, 'Deserts' were classified as being only hot and arid or semi-arid, a definition still used by many popular dictionaries, and applied to regions like the African Sahara Desert and internal regions within Australia. However, deserts can be cold as the fundamental definition of a desert is an area which receives less that 25 cm of precipitation annually, making the two largest deserts on Earth the Antarctica and the Arctic [114, 115]. Moreover, the Köppen system classifies 'oceanic' climates which are not in the open oceans but on land and in locations which have cooler summers than their counterparts at the same latitude, but do not have cold winters. The United Kingdom, Western Europe and New Zealand experience oceanic climates [116].

The highest point on land above sea level, Mount Everest, is still below the tropopause although the mountain is 10.5 times higher than the world's average land elevation of 840 m [117]. As air flows over land, friction between it and the land surface significantly influences the characteristics of the lower levels of the troposphere, but only to a certain level known as a 'capping inversion' [118]. Physicists and Engineers know that when a fluid flows over an object, such as a pipe wall, that a 'boundary layer' is formed from the object encumbering the flow until layer achieves a specific thickness. In a similar manner, air flowing over land obstacles form a boundary layer starting at the surface, which is often accompanied by vigorous turbulence. This layer is referred to as the Atmospheric Boundary Layer (ABL), or the mixed layer, and it has an average elevation (thickness) of about 2 km throughout a daily cycle varying from 200 m to 4 km, Fig. 1.12 [118].

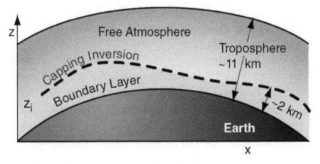

Figure 1.12 Atmospheric Boundary Layer [118].

This layer is where almost all of humanity lives and where almost all anthropogenic activities, such as agriculture, commerce and pollutant emission take place. In fact, about 80% of the global human population live below the average elevation with 50% living below 165 m [119]. The ABL thus plays a critical role in shaping global atmospheric circulations experienced by almost all living things and the resulting daily weather and longer-term climate conditions. As illustrated on Fig. 1.12, above the ABL the remainder of the troposphere is called the Free

Atmosphere where turbulence usually has little or no impact. The interaction between the ABL and its marine equivalent the Ocean Boundary Layer [120], as previously mentioned is "...*fundamental to the motions of both* [92]. The Ocean Boundary Layer (OBL) is discussed in the latter part of this chapter.

Although new scientific findings are constantly being made and insightful explanations forthcoming, the complexities of 'Atmospheric Circulations' are far from being entirely understood. Consequently, the descriptions of these circulations given here must be considered minimal overviews meant to assist a complete insight into the roles played by Oceans in weather and climate structures. Far more comprehensive publications are available which deal with meteorology and atmospheric circulations, probably none more so than the monumental textbook by Roland Stull [118].

1.3 The Great Ocean Conveyor and the Water Cycle

1.3.1 The Great Ocean Conveyor

As discussed in Sections 1.1 and 1.2, ocean water circulation is driven by wind generated surface currents and slower moving thermohaline currents in deeper waters. The two circulation sources are not wholly independent. A modern rendition of global thermohaline circulation has already been provided in this chapter in Fig. 1.8, but a circulation diagram published in 1987, again in 1991, and which has been repeated many times since, is that due to W. S. Broecker and J. L. Monnier [121]. Ironically, according to Broecker, the diagram was designed as a cartoon "*to help the largely lay readership of this* {Natural History} *magazine to comprehend one of the elements of the deep sea's circulation system*" [121]. This simplistic diagram has been almost indispensable to students and lay readers ever since. A comparison of the original 'cartoon-style' black and white diagram and a slightly more accurate color version used as courseware [77] is illustrated in Fig. 1.13.

Descriptions of thermohaline circulations appeared decades before Broecker's 1987 paper. The very distinguished oceanographer

H.M. Stommel published a series of papers starting in 1948 on the fluid dynamics of ocean circulations, which provided critical insights into explaining how and why ocean currents, such as the Gulf Stream, formed [122, 123]. Broecker used such earlier works to offer an explanation for the abrupt global climate changes that took place in a geological time period known as the 'Younger Dryas' some 13 to 11 millennia ago, when the global climate was in the process of moving from a cold glacial world to a warmer interglacial state [124], i.e., the end of the last ice age. He reasoned that the cause of the changes was that global thermohaline circulations were interconnected and had been suddenly halted. The name given to the linked circulations was the *Great Ocean Conveyor*, which, starting in the Norwegian sea, carries incredible amounts of water, heat energy, and salt around the globe before returning to its start point, as shown in Figs. 1.8 and 1.13, a journey that can take a millennium [125]. The conveyor acts as a major distributor of the heat energy received from solar radiation without which the poles would become increasingly colder and the air temperature in equatorial regions increasingly hotter, with the potential to make both regions largely uninhabitable. If this happened rapidly, climate generated human migrations could be overwhelming and mass species extinctions inevitable.

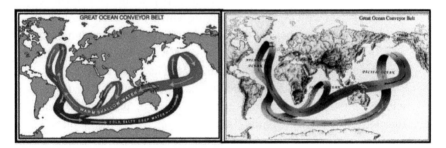

Figure 1.13 Broecker's Great Ocean Conveyor Belt: (left) colorized [77]; (right) original [121].

Moreover, without the conveyor, geographical locations at the same global latitude would largely have the same mean atmospheric temperatures. However, because of the existence of the conveyor, with its cold and warm ocean currents, places located at the same latitude are not guaranteed identical annual or monthly average air temperatures. The scale of the conveyor's influence does depend on a region's distance from the coast, the nature of ocean current, and other factors including the prevailing wind directions, elevation, and physical obstacles such as the orographic effect of mountain ranges.

1.3.1.1 The ocean boundary layer

The ocean conveyor also plays a role in the formation of the Ocean Boundary Layer (OBL) the marine equivalent of the Atmospheric Boundary layer. This is partially due, especially near coastlines, to the processes of (a) 'upwelling', mentioned in Section 1.2.1, by which winds blowing across the ocean surface push coastal water away and it is replaced by colder deeper water coming up to the surface, and (b) 'downwelling' which occurs when wind causes surface water to build up along a coastline and the surface water eventually sinks toward the bottom [126]. Away from coastal areas surface winds still play a significant role in the formation of the OBL, which has a variable thickness but can extend to a few tens of meters below the ocean surface [127]. The OBL thickness also has a daily and seasonal component due to several interactions such as (a) the radiative heating effects at different latitudes, (b) a limited form of upwelling and downwelling occurring within the OBL known as Langmuir circulations, (c) surface waves, (d) sea-ice dynamics, and (e) various chemical exchanges between the OBL and the troposphere as illustrated schematically in Fig. 1.14 [128].

Overall, the physical, biological, and chemical processes within the OBL, and at the sea-air interface, particularly with the ABL, are highly complex and continue to be the subject of intense ocean and atmospheric science research. It is known that the ABL–OBL interactions play key roles in weather and climate regulation and may be effected by climate change, but equally may also affect climate change, however, a comprehensive understanding of the many process remains elusive. For example, the roles of the chemical exchanges between ocean and atmosphere and how sea-ice alters the OBL are challenging research areas [129, 130]. Albeit while weather and climate are of profound significance to life on earth, how our fresh water supply is impacted by the systemic exchanges between atmosphere, land, and ocean is also of paramount importance. Yet, on a planetary scale it would be impossible to run out of drinkable water because of the *water cycle* which recycles our global water stocks by enabling the transitions from a vapour (gas) to a liquid to ice (solid), the three states of matter.

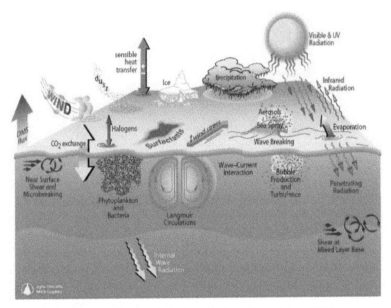

Figure 1.14 Factors effecting the mixed (or boundary) layer [128].

The main problem regarding potable water is not the amount, but the variation in geographical accessibility, a situation that is being exacerbated as global population rises are leading to consumption increases without an accompanying growth in local and regional water resources. It has been estimated that 1.1 billion people do not have access to safe drinking water and that 2.7 billion experience water shortages for at least one month every year, a figure that could rise to over 5 billion by 2025 [131]. This is of interest because changes in weather patterns can cause alterations to precipitation rates so that some regions experience devastating floods or equally harmful droughts. As the atmospheric and oceanic circulations generate regional and global weather patterns then any disruptions to these systems, natural or human, will impact accessibility to water, not only to drink, but also to support the agriculture on which humanity equally relies [132]. This impact will be manifest in changes to the natural water cycle[p].

1.3.2 The Water Cycle

The water cycle describes the routes taken by all water as it moves continuously around the globe. Often represented as a simple three-process system of evaporation-condensation-precipitation, the cycle is considerably more complicated, especially on land, involving at least nine major physical processes whose interconnections are yet to be wholly comprehended [133, 134]. Although, intuitively, the water begins and ends its journey in the oceans [133], as the cycle is a continuum there is no real start point and the cycle can be studied by considering the start to be any of the involved processes [134]. In global terms, all the processes take place simultaneously, spread over different geographical locations, with the most important of these processes being evaporation, transpiration, condensation, precipitation, and surface run-off as illustrated in Fig. 1.15 [133]. Evaporation from oceans and inland waters and transpiration from vegetation produces water vapour which, on ascending through the troposphere, will cool and revert to a liquid in the form of clouds. These will eventually return the water by precipitation (1) in the form of rain and snow, which will fall either directly back to the ocean or, with the bulk, onto inland and coastal areas, or (b) at the poles in the form of snow. Some of the water falling on land will run-off directly into the ocean by various means, while some *infiltrates* into the soil and may *percolate* into streams and fissures eventually becoming groundwater run-off [133] as also shown in Fig. 1.15.

---

[p]Usually physical geographers refer to this as the Hydrologic Cycle.

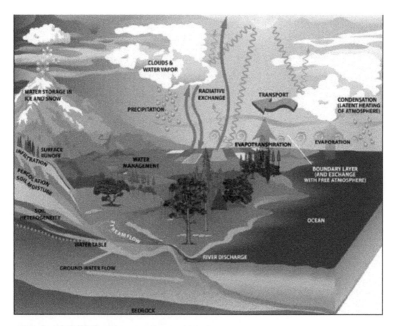

Figure 1.15 The global water cycle [133].

At the start of this chapter it was stated that there have been oceans on planet Earth for some 3–4.5 billion years [11] and it is estimated that they were as much as 26% more voluminous at these earlier times than they are today [135], but it seems likely that the amount of water in the water cycle has been sensibly constant for tens of millions of years; nonetheless, it can change by a minuscule amount due to various effects like volcanic activity. However, the water's distribution among the various physical processes is continually changing, often because of geographical location as shown on Fig. 1.16 [118]. For precipitation and evaporation, these process exchanges between the ocean and the atmosphere account for more than 75% of the water cycle [133, 136]. Clearly the water cycle is closely interrelated with the thermal energy exchanges between the atmospheric circulations, the ocean conveyor, and the orographic effects of land obstacles, and these interrelationships govern the Earth's climate and its natural climate variability.

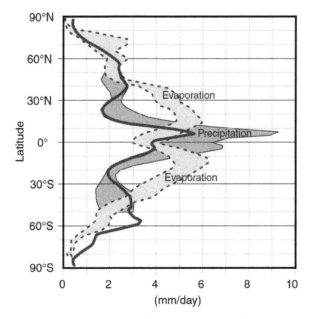

Figure 1.16 Average variation of precipitation and evaporation with latitude. *Note: The red line is an alternative estimate of precipitation distribution* [118].

1.4 Weather and Climate

1.4.1 General Effect on Weather of Ocean-Atmosphere Circulations

The combined effects of the ocean conveyor and the atmospheric circulations produce what can best be described as chaotic air motions, which are more commonly referred to as 'weather'. The usual definition of weather, found in numerous dictionaries, is the "state of the atmosphere at a particular time and place". Interest in weather is normally concerned with what the air temperature is going to be the next day, will it rain or snow, and how windy is it going to be? However, at any time and place, the weather will be primarily due to the prevailing air temperature, density, and pressure compared to adjacent areas or regions. While it is the temperature differences that cause winds, the strength of the wind, and its direction, are due to the pressure differentials which in turn are generated by the temperature and density differences. In areas of high surface temperatures, the warmed air will rise, usually creating low pressures, whereas colder air descends and generates high-pressure regions. Like other gases, air flows from high-pressure to low-pressure zones promoting winds.

The three-cell atmospheric system produces semi-permanent global areas of low pressure and high pressure. These low-pressure areas straddle the equator and are at higher latitudes well above the trade winds in both hemispheres. As the SST is higher in the equatorial regions significant water evaporation takes place and is accompanied by heavy rainfalls in the first portion of the Hadley Cell approximately 10°N to 10°S, which is why the largest land areas of rainforests are located in this latitude band. As the air circulation in the latter portion of Hadley cell, about 20°–30° North and South, starts to descend a high-pressure region is formed producing sensibly cloudless skies and much reduced rainfall. In these regions are the 'hot' deserts of Africa and Australia. The descending and ascending circulations within the other two cells also form areas of high and low pressure with similar, but not identical, characteristics to those of the Hadley cell pressure areas. At the poles, high-pressure areas experience little or no precipitation and are 'cold' deserts. Low regions are generated as the Ferrel Cell circulations approach the start of the polar cells causing an increase in precipitation and wetter climates, such as that of the United Kingdom [e.g., 137].

The foregoing description can be helpful in appreciating the pressure layering effect of the basic three-cell system and the resulting weather patterns, but it has to be realized that the precise location of the layers does change, especially seasonally, and the weather patterns within the layers are modified by the hot and cold effects of various current streams that comprise the ocean conveyor. For example, the US cities of Omaha, Nebraska, and Eureka, in the sunshine-state of California, are both located at the 41°N latitude, but Omaha's highest monthly temperature is twice that of Eureka, whereas the lowest Omaha temperature is, at about −6°C, well below Eureka's of +8°C. Omaha is over 2000 km from any sea whereas Eureka experiences the cold Californian offshore current circulation. The conveyor effects are particularly conspicuous in the Western European winter season where the warming impact of the Gulf Stream is pronounced. At 65°N, the average January temperature in Brønnøysund, Norway, is almost 15°C warmer than that of Nome, Alaska [138] and in the same month, Bergen, Norway is almost 7°C warmer than Halifax, Nova Scotia which is 1700 km closer to the equator [73]. Arguably these phenomena are part of the 'normal climate' —in the absence of any climate change disruptions. Also, part of the normal global and regional climates are the seasonal strong wind systems such as monsoons and cyclones. However, the formation of such systems is not the same, although both involve seasonal differences between ocean and land surface temperatures and pressures.

1.4.2 Monsoons

When talking about 'monsoons', the persistent general perception is one of massive rainfalls, but monsoons are wind systems which reverse direction between the summer and winter from what are known as 'sea-breezes' to 'land breezes' [126]. However, this type of reversal is not unique to monsoons and can occur on an almost daily basis near to coasts when there are measurable temperature and pressure differences between the sea and land. These frequent phenomena, which are also seasonal and location dependent, are the result of solar radiation heating up the land quicker than the sea, because of their different specific heat capacities. As the air ascends, and generates a low-pressure area, a horizontal flow of air from the sea to the land is generated. At night, in the absence of sunlight, the land cools more rapidly and, if the ocean is warmer than the land, the airflow will be from land to sea. Monsoons exhibit the same sea breeze—land breeze reversals but over a season rather than a day, and on a 'continental' scale, carrying large quantities of water vapor [118]. In the summer season the land is hotter and the consequent formation of low-pressure areas cause winds to blow from the oceans to land where the water falls as heavy precipitation. In the winter the lower pressures are over the warmer oceans and the winds flow from land to ocean.

The most well-known monsoon is the Indian Ocean monsoon, which is part of the larger Asian monsoon system. As mentioned earlier, the English polymath Edmund Halley published a map of the Indian monsoon winds as early at the latter part of 17th century [100]. However, observations of the Asian system, both the wet and dry phases date back 3 to 4 millennia [139]. In a 'good-wet' year monsoons result in large amounts of rain falling on the land, for example the average rainfall in Cherrapunji, India, at 25°N, considered to be one of the wettest regions on Earth, is over 10.3 m and has a recorded maximum of 26.5 m [140]. For comparison, the United States wettest city in New Orleans with an annual average rainfall of about 1.6 m [141], while major cities like London, UK and Toronto, Canada, average 0.6 m and just under 0.8 m, respectively. Although the population of India is now approaching 1.4 billion, the regions which experience the larger Asian Monsoon system are home to almost 4.7 billion or 60% of the global population. The monsoons are crucial to the economic well-being of the countries in this region, as the rainfalls sustain their agricultural activities without which there would be catastrophic crop failures [139].

As more data become available on monsoons, especially now that satellite measurements are commonplace, the formal definition of what constitutes a monsoon is changing [142]. The term increasingly refers to regions where there is an alternation between winter dry, and summer rainy, seasons and a monsoon domain is defined as a region *"where the local summer-minus-winter precipitation rate exceeds 2.5 mm/day and the local summer precipitation exceeds 55% of the annual total"* [143]. Hence, monsoons are not phenomena solely confined to the India-Asia region [140], but now include the North and South American monsoons, the Australian monsoon, the West African monsoons, and many others, all of which can be classified as one of three main types, tropical, sub-tropical, and temperate-frigid as illustrated in Fig. 1.17 [139]. Not surprisingly, there are advocates for the concept of a *global monsoon system* [144].

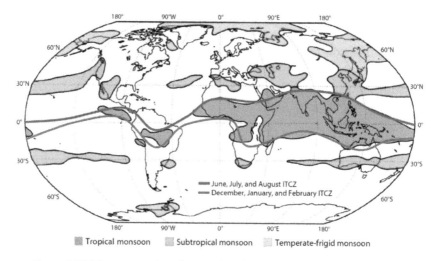

Figure 1.17 Monsoon regions by type [139].

Almost 400 million people live in the West African region[q], but average rainfall varies significantly from the latitudes closer to the equator, 5°N, at 1.6 m, to the more northern boundary at about 18°N where the rainfall is only 16.5 cm. These averages can be affected by annual variabilities, which can increase rainfall from 10% in coastal areas to as much as 40% inland [145]. Many areas have agriculturally based economies and in rain-poor years the lack of rainfall, like in the Asian regions, can have a devastating impact on the well-being of the involved populations [146]. The West African monsoons also play a pivotal role in the formation of clockwise tropical storms which, as they are blown westward by the trade winds, can become hurricanes[r]. These storms gain energy from both the warm ocean surface and the ascending warm moist air. On reaching land the hurricanes quickly abate as they have no access to the ocean energy drive, but until they do they can cause overwhelming damage in coastal areas because of excessive sustained wind speeds, between 119 km/h and 252 km/h or higher, massive short-term rainfall, and a *storm surge* caused by the winds literally pushing water on shore creating extraordinarily high

[q]As defined by the United Nations.
[r]Also known as typhoons and cyclones depending upon their geographical location.

sea levels [126, 147, 148]. The prerequisites for such occurrences are the development of thunderstorms off the coast and sea surface temperatures of at least 26.5°C. If climate change results in rising sea surface temperatures in areas prone to hurricane formations, then the likelihood of these hurricanes being more frequent and stronger is very real, as is the lengthening of the average hurricane season [149].

1.4.3   Oscillations

The weather effects described in Sections 1.4.1 and 1.4.2 contribute to what can be described as the normal climate although short-term disturbances, called anomalies, may occur [118]. When such anomalies are reasonably regular, longer lasting, occurrences they are known as *oscillations* of which many have been identified [150]. These events are chiefly characterized by changes in air pressure, surface sea temperature, and prevailing wind direction over the oceans from those present under normal conditions and although they are located regionally, they can also impact global weather patterns. One of the most important oscillations is the El Niño-Southern Oscillation (ENSO), which occurs in the equatorial region of the Pacific Ocean between Australia and Peru. El Niño is an ocean oscillation caused by changes in SST, whilst the Southern Oscillation is caused by changes in sea-level pressure; these two phenomena combine to form ENSO [118]. El Niño occurs when the SST in the region is measurably higher than normal, which occurs in a cycle on average between 2 and 7 years, and its effects can last for up to a year and sometimes longer. If, however, the SST is significantly lower than the normal average then the oscillation is known as La Niña and its weather impact is basically the reverse of those associated with El Niño and occur in a similar cyclic manner. Both the warmer and cooler oscillations have their most noticeable effects on weather during both the southern hemisphere's and the northern hemisphere's winter, peaking usually between December and February in the North and June to August in the South, as illustrated in Fig. 1.18 [151]. It can be seen on the top left hand side of diagram, that in the calendar period around Christmas, the waters of the Western South American coast are warm, which is not the normal condition. Fishermen in the 17th century noted that this phenomenon occurred occasionally off the Peruvian coast and gave it the name El Niño [152]. However, it was not until the 20th century that an atmospheric circulation pattern known as the 'Walker Circulation' was recognized as being interconnected with El Niño to produce the EN-Southern Oscillation and that a similar, but opposite effect, La Niña, occurred when the seasonal SST off the equatorial region of South America was detectably below the normal condition [152, 153].

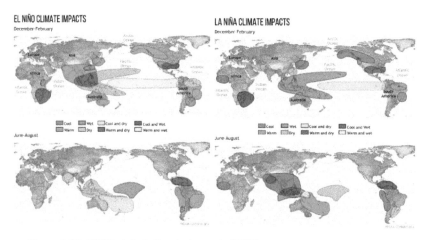

Figure 1.18 ENSO global climate impacts [151].

The Walker Circulation, or convection cell, is a West-East event rather than the South-North circulation of the three-cells discussed in Section 1.2.3.2. For this distinct circulation, trade winds flowing over the ocean surface towards the west from South American create an expanse of warm water close to Indonesia and Australia. This warm water interacts with the overlying atmosphere generating ascending air and precipitation, synonymous with a low-pressure region. As this air reaches higher altitude it flows west to east and descends when it encounters the high-pressure region close to the South American coast forming a cell or loop. Under normal conditions, or neutral ENSO conditions, the circulation has little or no climatic influence on Australia but brings rain to parts of Melanesia, especially the world's second largest island—New Guinea and the Solomon Islands, as shown in Fig. 1.19 [118, 154].

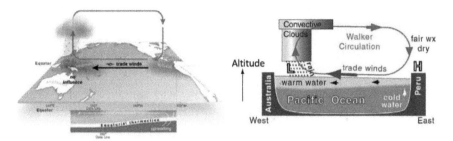

Figure 1.19 Normal or neutral ENSO characteristics [118, 154].

During an El Niño episode the high-pressure system over the eastern Pacific fades and the trade winds become weakened, sometimes reversed, and the normal areas of warmer surface water, rather than forming close to Indonesia and Australia, are confined to the central and eastern Pacific. This can lead to water temperatures increasing in South American coastal areas either side of the equator by as much as 8°C [126], the conditions observed by the Peruvian fisherman over two hundred years ago. For eastern Australia the winter rainfall can be significantly reduced and may lead to droughts and increased bushfire incidents. Largely opposite effects are experienced in La Niño events when uncharacteristically cold water regions form off the South American coast due to increased upwelling and the trade winds become stronger, pushing far greater amounts of warm surface waters towards Indonesia and Australia bringing times of intense rainfalls, higher risks of widespread flooding, and more storms to most of Australia [154]. However, as shown in Fig. 1.18, El Niño and La Niña result in weather impacts globally and not just in the narrow equatorial region of the Pacific. This is mainly due to the wide-ranging coupling between atmospheric-ocean interactions associated with ENSO but prevalent outside the ENSO region, Fig. 1.20 [155]. It should also be noted that ENSO, both in the El Niño and La Niña phases, can impact the formation of sea ice in the Antarctic regions, but not immediately.

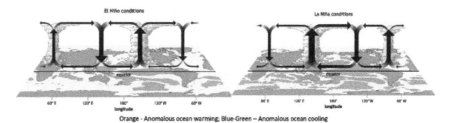

Orange - Anomalous ocean warming; Blue-Green – Anomalous ocean cooling

Figure 1.20 Differing ocean cooling and warming pattern with El Niño and La Niña ENSOs [155].

As stated earlier, the physical motions of the oceans and atmosphere are intertwined in a highly complex manner creating changing weather systems that are difficult to forecast with any high degree of certainty [10]. Nevertheless, early warnings of the onset of ENSO, or the formation of Hurricanes, are critically important, such that many national governments like those of the United States, Australia, and the Philippines operate climate monitoring centers to track such events. The World Meteorological Organization (WMO), through its world weather information service, collects data and warnings from its 170 members to produce daily global forecasts as well as severe weather warnings [156].

1.4.4 Other Weather Changing Factors

To paraphrase Roland Stull: "Overall, the Earth's climate is remarkably steady" [157]. However, in terms of weather this situation is somewhat different. While the 'seasons' are invariably present, with summers being relatively warm and winters relatively cold, in the Northern Hemisphere, and the opposite in the Southern Hemisphere, the short to medium term effects, such as the 'oscillations', occasional volcanic eruptions, and periodic alterations in solar activity, can dramatically alter those seasonal weather patterns. The ENSO discussion provided examples of how oscillations can impact weather and, while it is, arguably, the foremost oscillation in creating global weather changes, other oscillations are constantly being discovered with cyclic occurrences varying, from tens of days to tens of years, but their effects tend to be restricted to particular geographical regions or zones [70, 118, 126]. The impact of changes in the amount of solar radiation reaching the Earth and the frequency of volcanic activity is far more difficult to predict.

Probably because of the growing concerns about the melting of large glaciers and the polar ice-caps, there has been far more interest this century in the formation and characteristics of the Arctic and Antarctic oscillations, including the Air-Ice-Ocean interactions in these polar regions [158]. Observations, and human in-situ data collection, are especially difficult in these regions, but with the increasing use of satellites, scientists can gain insights into the how's and why's of changing cryosphere features, including sea ice extents and their influence on circulations, by comparing prevailing conditions to long-term averages [159]. Nevertheless, the paucity of these data have required scientists to develop models to 'fill-in-the knowledge gaps' in the polar cryosphere, but they can only be validated if more data become available, a somewhat 'catch-22' situation. Coupled with all the complexities of weather generation and forecasting, and the ocean's roles in these intricacies, the possible effects of climate change also need to be taken into account. But what is climate change?

1.5   Oceans and Climate Change

In 1992, the United Nations, in their Framework Convention on Climate Change (UNFCC), defined what climate change means, *"a change of climate which is attributed directly or indirectly to human activity that alters the composition of the global atmosphere and which is in addition to natural climate variability observed over comparable time periods,"* [160]. So, when discussing 'Climate Change', it is a tad specious to espouse that it is caused by humans when that is precisely how 'Climate Change' is defined! This had led to those suggesting that *"climate had always changed"* being labelled as 'climate deniers' or 'climate skeptics' which is unfortunate, and likely a case of misinterpretation, since persistent natural weather changes are now defined by the UNFCC as *climate variability*. Without an awareness of the different definitions, discussions can often be at cross-purposes and not helpful. Consequently, it can be argued, that any climate debates should be not so much about 'if' changes in climate occur, but what causes these changes. But what is climate?

The scientific community generally defines climate as the average weather, in terms of precipitation, prevailing winds, flooding, heat waves and so on, at a specific location or region, over a 30-year period. There are two main reasons for using 30 years; (a) The WMO instructed its members almost a century ago to use this timeframe to calculate 'Climate Normals', and (b) a general rule in statistics is that at least 30 numbers are required to determine a reliable estimate of their mean or average [161]. The '30-year rule' is then a convention rather than a proven scientific standard and it follows that it is not, *"the only logical or 'right' way to define a Climate Normal"* [161]. Nevertheless, the definitions of climate change and the time period are implicitly, and sometimes explicitly, embodied in the bulk of written and digital publications, whether scientific, political, or otherwise. This needs to be appreciated when examining climate change matters.

The core argument surrounding anthropogenic climate change is that the ill-named Greenhouse Effect [118], necessary to maintain comfortable air temperatures over the habitable regions of Earth, has been negatively impacted by the increasing amounts of anthropogenic Greenhouse Gases (GHG) such as Methane, Nitrous Oxide, and especially Carbon Dioxide generated since the advent of the 18th century's First Industrial revolution. By negative, it is not meant that the effect has been lessened, but rather than it has been increased, so that more of the solar radiation reflected from the Earth's surface back into the atmosphere is now being returned again to the surface, resulting in both elevated average air temperatures over land, and water temperatures in the oceans. These increases could lead to a variety of unwelcome impacts on local and global weather patterns and the climate.

1.5.1   Abrupt Climate Change

How do we know that Climate Change is impacting, and has impacted, oceans? What are the measurable or, at least, observable indicators? The United States Environmental Protection Agency (USEPA) identified several ocean properties that could be used as indicators of climate change alterations, including sea level, ocean acidity, and sea surface temperature [162]. To these indicators could be added those related to 'abrupt climate change', a significant topic of discussion in the early 21st century when scientists, examining historic and fossil evidence, discovered that the Global climate had abruptly changed, i.e., within a decade, on several occasions [163]. Could such an event happen again? Yes, according to a report by the US National Academy of Sciences (NAS) which stated, "*available evidence suggests that abrupt climate changes are not only possible but likely in the future, potentially with large impacts on ecosystems and societies*" [164]. The historic, abrupt changes were linked to the Great Ocean Conveyor being stopped, or becoming significantly slower, particularly its Atlantic component [124, 165].

As noted in Section 1.3.1, the historical evidence is associated with a geological period, the 'Younger Dryas', which occurred a few millennia after the last ice age and lasted for several centuries [125]. The Northern Hemisphere did get colder and dryer although elsewhere the climate changes were variable including warming in some regions. While the global population during this time was 20 times less than it is at present, it has been estimated that, during the initial period of the Younger Dryas, the population declined by about 50% [166]. The warnings about the growing onset of 'Global Warming' in the 1980s led to concerns among some climate scientists that this would result in more freshwater entering the North Atlantic due to ice and glacial melting triggered by increased air and sea temperatures. This influx would prevent the normally salty Atlantic waters from sinking, which would halt the heat being carried by the Gulf Stream entering the northern waters, and stop the conveyor. This concern was heightened when observations made in 2005 by a team at the UK's National Oceanographic Centre found that the circulation speed of the conveyor was reducing. However, later observations by the same team found that the conveyor speed varied considerably over relatively short time periods with no clear trend [167]. The precise linkages between the ocean conveyor and abrupt climate change remain unresolved, but researchers have continued their quests to collect evidence of such changes and identify the causes [e.g., 168, 169].

1.5.2 Sea Level Changes

The average global sea level has been rising since the start of the 20th century. With more than 600 million people living in coastal areas that are less than 10 m above sea level, and almost 2.4 billion living within 100 km of a coast [170] the rise is a cause for concern. The recent IPCC report [2] on oceans contains model projections using their Representative Concentration Pathway (RCP) scenarios and the more pessimistic of these suggest that sea levels could rise by as much as 0.63 m by the end of the 21st century, and 6.6 m by 2500, almost five centuries from now [2, 171]. However, an often-quoted scenario is by how much would the sea level rise if all the glaciers and ice sheets melted [e.g. 172]. Such an occurrence is improbable, but various national agencies have estimated that if it did happen sea levels could rise between 60 and 70 m above current levels. This could suggest that the present IPCC estimates are perhaps conservative [126]. However, since it has taken over 24,000 years for the post-glacial sea level rise to reach 125 m, half the total ice melt rise scenario, is there any need to be concerned by possible sea level rises that, though unlikely, may occur five centuries to perhaps 10 millennia from now? Probably not, but as sea levels rise there needs to be an awareness that a modest rise of between 0.5 and 2 m could displace 1–2 million people in low-islands in the Caribbean, the Indian, and Pacific oceans. Already, as many as 21.5 million people have been displaced annually since 2008, not especially as a result of rising sea levels, but by sudden weather events considered to be due to ocean warming [170]. This awareness is epitomised by the quotes attributed to Helen Fricker, a glaciologist at Scripps Institution of Oceanography that, *"Nobody's debating that sea-level rise is happening. It's back to how much, how fast, … It's healthy to have this debate"* [173].

However, to understand the nuances and results of such debates it is important for the non-experts to appreciate what is meant by 'sea-level' and this is not as straightforward as may be imagined. Part of the difficulty lies in defining a datum from which sea level can be measured. The center of the Earth may seem a good starting point, but the Earth's ellipsoid shape is somewhat flattened at the poles compared to the equator, which means that the same 'sea-level' measured from the center to equator would be different from that measured at the poles by 42 km. The effects of gravity on 'sea-level' can be even more pronounced as a result of 'Newton's Law of Universal Gravitation', which basically states that the gravitational force, or attraction, is stronger between two bodies when the masses are greater, and when they are closer to each other. When applied to the Earth's land and oceans masses this invariably means that, at locations where the Earth's gravitational forces are at their strongest, the mean sea level will be higher, and where the forces are weaker, the mean sea level will be lower [174]. As measured by modern satellite altimetry the spatial variation of mean sea level can be as much as 100 m [175].

To mathematically overcome the flattening issue, scientists modelled (approximated) the shape of Earth by using a defined oblate spheroid, known as the *reference ellipsoid.* This model was significantly enhanced in the 1990s by including more, and finer resolution surface gravity data, and became known as Earth Gravitational Model 1996, or EGM96 [176]. This was further improved in 2008 by EGM2008 [177] and annual continuous improvements have been made since 2014. These models can be used to calculate the 'Geoid' which describes the irregular shape of the earth and should be the true zero surface for measuring elevations and, hence, mean sea levels [178]. However, although the Geoid is a good approximation there are problems partly due to further gravitational irregularities being constantly discovered and national variations in defining zero elevation. Moreover, the differing meanings of the same terminology, such as 'dynamic sea level', as used by various scientific disciplines, can lead to confusion and ambiguities, especially in climate science literature [179].

There is no doubt that sea-level science is a complex and challenging topic, but as political concerns about future sea-level rise it could lead to costly global, regional, and local policy decisions, attempting to find solutions to non-existing problems, it is important to have as much accurate physical data as possible so that universal models can be fully calibrated and validated. Regardless of the present vagaries of modelling potential climate change effects on sea-levels, because of the obvious implications, we still need to know if mean sea-levels are rising and, if so, for how long has the rate been accelerating? Are measurements available that can help to provide answers or at least unequivocal insights to these questions? Fortunately, a considerable body of such data continues to be collected.

The measurement of sea levels and tides has a very long history using some type of vertical tide-gauges. Such instrumentation is still in use and for over the last 150 years has been employed at many global locations with the data being recorded and collated [180, 181]. However, for a variety of reasons data collected from tide gauges are defined as local relative sea level (RSL), a combination of sea level rise and the vertical land motion [182]. Then once again models are used to calculate a Global Mean Sea Level (GMSL), but since 1992 satellite altimetry has also been used to measure global sea-level trends [182]. The extensive website and links associated with NOAA Tides & Currents [182] list many co-operating tidal stations nationally and globally with many decades of data being made available. The trends indicate that, in some locations and regions, sea-levels are rising whilst in some they are falling; some exemplars are shown for RSL tidal gauges together with the satellite global mean in Fig. 1.21 [181]. As of March 2020, according to NASA satellite data global sea levels are rising at an annual rate of 3.3 ± 0.4 mm, slightly higher than shown on Fig. 1.21 and twice the rate recorded in the 20th century prior to the advent of satellite altimetry.

In 2017, the US Inter-agency Sea Level Rise Taskforce reported that depending upon the level of future GHG emissions, sea levels would rise, from the start to the end of the 21st century, by as little as 0.3 m to as much as 2.5 m [183]. The lower limit is based on current satellite data citied in the previous paragraph. If exemplar RSL trends are compared with the most recent mean sea level (MSL) rise rate from satellite data for the second half of this century, it can be seen, Fig. 1.22, that at Sewell's point on the North-East coast of the United States sea level rises above the global MSL will be experienced, whilst other global locations, such as the Swedish city of Stockholm, will continue to encounter falling sea levels. By the end of this century the RSL difference between Sewell's Point and Stockholm will be almost 70 cm, but still far less than the RSL difference between the falling levels at Skagway, Alaska and rising levels at Chomklao, Thailand, of 2.78 m. So, as to the question, are sea-levels rising, the answer from the trending data is yes, an affirmation of Fricker's statement. The rates have been accelerating in the first two decades of this century, but whether they will continue to accelerate remains a matter of conjecture. Globally the mean sea level is rising, but there are coastal areas where the level is significantly falling, as well as other areas of sizeable increases. Can these differences be fully explained by Climate Change caused by increasing levels of GHG? As the sea level changes appear to be regional then can it be supposed that Climate Change is regional?

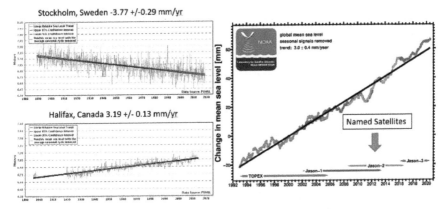

Figure 1.21 RSL exemplars and satellite global means [181].

Not as such, but it is perhaps worth noting that the RSL rates in close proximity to the internationally accepted CO_2 measurement facility in the Southern Hemisphere, Cape Grimm,

Tasmania, at between 0.38 and 0.75 mm/year, are well below the Global Mean Sea Level (GMSL) rates of 3.3 mm/year. In the Northern Hemisphere CO_2 measurements are taken at Mauna Loa Observatory on Hawaii's Big Island, but the RSL rates for all the Hawaiian tidal gauge stations are below the GMSL, varying from 1.51 to 3.12 mm/year [182]. However, rising sea levels are mainly caused by meltwater from glaciers and ice sheets combined with the thermal expansion of seawater as SSTs increase because of Global Warming, a key part of Climate Change [183]. During the last ice age, both Tasmania, once part of Antarctica, and the Hawaiian Islands were covered by thick ice, but these melted millennia ago. Thus, what at first sight may appear to be somewhat anomalous data are, arguably, merely indicative of one aspect of Climate Change on the oceans.

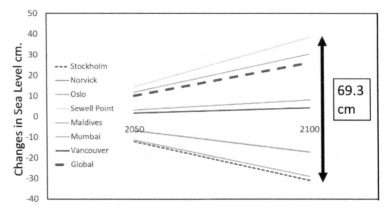

Figure 1.22 Exemplar sea level changes 2050 to 2100 (author generated) based on 2020 data [181].

1.5.3 Ocean Acidification

The oceans are a natural carbon sink for absorbing and storing atmospheric carbon; maybe storing as much as 50 times more carbon than the atmosphere. There are two basic mechanisms involved in the processes, (a) the 'biological pump' and (b) the 'physical pump'. The former transfers surface carbon towards the ocean floor via the food chains, whilst the latter is due to ocean circulation and is the least understand of the two pumps [184].

The oceans can both capture, and release, CO_2, the amounts are estimated to be in balance until the past three decades or so. The lower the sea temperature the easier it is for CO_2 to be absorbed, but as the amounts of atmospheric CO_2 are increasing, and are being accompanied by rising global average temperatures, there are concerns that the resulting climate change will adversely impact the ocean's ability to be a carbon sink, leading to abnormal levels of CO_2 being released from the ocean. Consequently, focused scientific investigations on the impact of carbon dioxide on ocean water chemistry have become more prevalent over the past 15 years, especially for high latitude regions [185], and it has become popular to talk about 'the other CO_2 Problem' [186].

The term commonly used by scientists to describe this carbon dioxide problem is 'Ocean Acidification'. This description may conjure up government and popular perceptions of oceans taking on the characteristics of electrochemical battery acid, but this is not the case as ocean water is still moderately alkaline based on the pH scale [187]. The pH scale has a range from 0 (very acidic) to 14 (very basic or alkaline), and is logarithmic, so a change of 2 units, say from 4 (e.g., tomato juice) to 2 (e.g., lemon juice), means an increase in acidity by a factor of 100. Numerically, battery acid has a pH of zero, distilled water is neutral with a pH of 7, and seawater now has a pH of 8.1, but over a century ago the pH was 8.2, although there are doubts about the accuracy of pH data collected prior to 1989 [188]. Regardless of exactly how much the pH of seawater has changed, the 'basicity' is known to be lower in this century than it was in the last. It should be noted that chemistry textbooks some three decades ago would signify the pH scale as a measure of 'acidity' or 'basicity', but it now is commonly referred to as an indication only the acidity of an aqueous solution. Thus, although seawater in chemical terms is still a 'base' it has become a convention to describe the lowering of the pH in seawater as 'Ocean acidification'. Perhaps, a description such as 'Reduced Ocean Basicity', while accurate, would likely create a more benign mindset attracting less interest or concern than 'Acidification'.

How does CO_2 become absorbed in the ocean and alter water chemistry characteristics? The CO_2 reacts with the seawater to produce a weak acid, called carbonic acid whose acidity, on average, is about the same as tomato juice and approximately 10,000 times less than battery acid. The carbonic acid then quickly dissociates becoming bicarbonate, carbonate, and hydrogen ions which, together, are described by ocean scientists as 'dissolved inorganic carbon' and play crucial roles in what is known as the 'Carbonate System' [189]. As a result of all these reactions, the amount of hydrogen ions increases, lowering the ocean's pH value and; at the same time, the availability of carbonate ions decreases. Why is this important? The carbonates combine with calcium in the seawater to provide shell and skeleton building materials used by some plants and animals to provide structure and protection. A declining proportion of carbonate ions can lead to a condition called 'undersaturation' which presents harmful challenges to the formation of shells and skeleton causing them to dissolve or deform. In some cases, species can be prevented from reaching their adult stage [190].

Undersaturation occurrences could severely affect the ocean food chain, especially 'capture' fisheries. However, since 2013, global seafood production by aquaculture has been greater than wild fish catches [191]. So, the possible negative impacts on human population food supply and economics are difficult to quantify, except that those coastal communities which rely on wild fish catches could be adversely affected. It should be noted that ocean scientists have also defined a seawater characteristic called 'alkalinity', a somewhat unfortunate choice as it can lead to some confusion among those not directly involved in the discipline believing that acidity and alkalinity are identical parameters. In addition to the carbonate ions present in seawater there are numerous other ions such as phosphates and silicates which chemically grab some of the hydrogen ions associated with pH levels, acting as 'buffers' against increasing ocean acidity, a measure of this effect is termed 'Alkalinity' [185, 189]. Scientists investigating the ocean's water chemistry determine OA (Ocean Acidity) and TA (Total Alkalinity) separately as part of climate change studies [185]. There remain many uncertainties in this relatively new area of scientific endeavour, but it has been observed that OA and TA vary by region especially in coastal areas where there are rivers flowing into the sea and in some locations where there are strong current circulations and air-sea interactions. Since water flow from land is largely responsible for bringing the salts into the ocean these variations are not surprising. However, such flows also lead to another environmental concern known as 'Eutrophication'.

1.5.3.1 Eutrophication

The Eutrophication problem occurs when an excess amount of nutrients, such as nitrogen and phosphate compounds, are delivered to a water body such as a lake or ocean estuary. The presence of these nutrients can be natural and are important for plant growth. However, as the global population has increased seven-fold since the start of the 19th century, human generated and animal waste is finding its way into freshwater, and eventually coastal sea areas, along with significant amounts of the fertilizers, pesticides, and insecticides used in terrestrial food production. These activities have caused excess inputs of nutrients resulting in an excessive growth of water-borne algae, often generating noxious algae blooms. The algae growth clouds the water, blocking the amount of solar radiation received below the surface, and, more importantly, as the algae decomposes oxygen levels in the water are reduced. These effects are the result of human activity and lifestyles, and to differentiate between natural and anthropogenic impacts the term 'Cultural Eutrophication' is used [192]. But what has this to do with ocean climate change? The main connection is that higher water temperatures play a role in increased algae growth [193]. Indirect effects include concerns that cultural eutrophication in lakes will increase methane GHG emissions into the atmosphere by between 30% and 90% by 2120 [194]. This too could cause further rises in SSTs and result in further changes in ocean water chemistry.

1.5.4 Sea Surface Temperatures

Sea Surface Temperature is a key parameter in all aspects of ocean characteristics from circulation patterns to air-sea interactions, sea levels, weather events caused by oscillations and monsoons, and regional climates. But what if the normal cyclic variability of SSTs were to be adversely impacted by climate change? Increases in SST would result in the thermal expansion of the oceans causing sea level rises and strongly influencing increases in ocean acidification with the all the consequential harmful effects previously addressed. But have average global SSTs been rising? The answer is yes for at least four or five decades [195]. Prior to the 1970s, although reliable, albeit sparse, data had been available for close to a century, the results showed periods of decreases and increases relative to the 1971–2000 average. Nevertheless, the average incremental increases in SSTs from 1901 to 2015 have been determined to be 0.07°C per decade for a total increase of 0.8°C [195]. However, using the same data, some scientists, but for a slightly different time period, have estimated that the rise is 0.13°C per decade [196]. Climate modelling using IPCC RCP scenarios have indicated that incremental increases could be twice to four times higher in the last quarter of this century likely resulting in mean global ocean temperatures increases of 1 to 4°C by 2100 [197, 198].

All that is really known is that ocean sea surface temperatures since the late 1970s have been rising, but between 1880 and about 1940 they were decreasing, then the increases of the 1940s quickly reversed for another 2–3 decades. If increased GHG emissions are causing the recent rises why have not they been manifest until recently? The main rationale for this occurrence is that increases in the solar radiation, direct and returned, by the 'Greenhouse effect' take time to warm the oceans, because of the large volume and surface area of the oceans and their high specific heat capacity. Whatever the reasoning, sea surface temperatures are rising but it has yet to be seen whether the increases are manageable, i.e., self-correcting or of such a scale that, as outlined in the previous sections, there will serious consequences for the Earth's ecosystems.

1.5.5 Overall Effects and Impacts

Some of the key indicators of ocean climate change and insights into how they are altering have been discussed in Sections 1.5.1 to 1.5.3, but a more inclusive synopsis of climate change effects provided by the IPCC [2] is illustrated diagrammatically in Fig. 1.23.

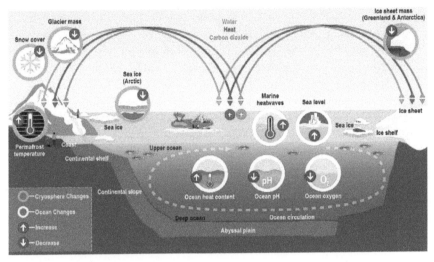

Figure 1.23 Changing ocean and cryosphere climate indicators [2].

1.6 Concluding Remarks

In this chapter the history of the oceans and their known characteristics, physical, and to a lesser extent, chemical, have been addressed, but with only enough space to portray the primary aspects of these complex almost intractable topics and to provide pathways for those interested in further and more advanced study. In doing so it has been emphasized that from the observations of ancient human to modern satellite ocean measurements the crucial roles played by the oceans in the life of the planet, its inhabitants, and their well-being are finally being revealed, but not as expeditiously as is necessary to tackle the legitimate concerns about the state of the ocean. Indeed, human knowledge of the ocean systems is far from complete and this ignorance is leading to severe penalties in terms of the loss of human, plant, and animal life, and infrastructure damage globally costing billions of dollars, which only the richer countries can afford to rehabilitate. Even in the coastal and low-lying regions of these 'well-off' countries the impact of Climate Change may prove overwhelming; the poorer regions, which rely heavily on the oceans for their sustenance, way of life, and economic well-being, could be irreparably ruined.

The oceans are vast and deep, but despite the impressive strides being made in the ocean sciences there are many uncertainties in the levels of understanding of the roles that oceans play in weather and climate, which could likely be exacerbated by the potential negative impacts of anthropogenic climate change. The need for more data is paramount if the models of our Earth systems are to be improved. These models can help rational decisions to be made in combating, not only the effects of climate change, but also climate variability. It may be that many of the climate concerns are ill-founded in terms of scale, but their impacts may also be underestimated, we just do not know.

References

1. NOAA, 2020. What Are El Niño and La Niña? National Ocean Service, 10 February. https://oceanservice.noaa.gov/facts/ninonina.html.

2. IPPC, 2019. The Ocean and Cryosphere in a Changing Climate, in *A Special Report of the Intergovernmental Panel on Climate Change*, Pörtner, H.-O., et al., eds., p. 765. https://www.ipcc.ch/site/assets/uploads/sites/3/2019/12/SROCC_FullReport_FINAL.pdf.

3. UNESCO, 2019. United Nations Decade of Ocean Science for Sustainable Development (2021–2030). https://en.unesco.org/ocean-decade.

4. UNFCC, 2020. Action on Climate and SDGs. https://u5.nfccc.int/topics/action-on-climate-and-sdgs/action-on-climate-and-sdgs.

5. UNESCO, 2017. *The Ocean We Need for the Future We Want: A Proposal*: p. 8. http://www.unesco.org/new/fileadmin/MULTIMEDIA/HQ/SC/pdf/IOC_Gatefold_Decade_SinglePanels_PRINT.pdf.

6. The Nippon Foundation, 2018 GEBCO Seabed 2030 Mission Statement; p. 6. https://www.gebco.net/documents/seabed2030_brochure.pdf.

7. Saul, J., 2020. Map of World's Uncharted Ocean Beds Takes Shape Despite Crisis. *Reuters*. 20 June. https://www.msn.com/en-gb/news/world/map-of-worlds-uncharted-ocean-beds-takes-shape-despite-crisis.

8. Fox-Kemper, B., et al., 2019. Challenges and Prospects in Ocean Circulation Models. *Front. Mar. Sci.*, February, vol. 6, no. 65, p. 29: doi:10.3389/fmars.2019.00065.

9. Fox-Kemper, B., 2018. Notions for the Motions of the Oceans. *New Frontiers in Operational Oceanography*. Retrieved from http://purl. flvc.org/fsu/fd/FSU_libsubv1_scholarship_submission_1536167323_ 6ef1328f.

10. Zhang, F., et al., 2019. What Is the Predictability Limit of Midlatitude Weather? *J. Atmospheric Sci. Am. Meteorol. Soc.*, April, vol. 76, pp. 1077–1091. https://journals.ametsoc.org/jas/article/76/4/1077/343664/ What-Is-the-Predictability-Limit-of-Midlatitude.

11. Cenedese, C., and Duxbury, A. C., 2020. Ocean, Earth Feature, *Encyclopædia Brittancia*. https://www.britannica.com/science/ ocean.

12. NOAA., 2019. What's the Difference between an Ocean and a Sea? https://oceanservice.noaa.gov/facts/oceanorsea.html.

13. NASA., 2020a. Living Ocean. https://science.nasa.gov/earth-science/ oceanography/living-ocean, accessed 18 March 2020.

14. International Chamber of Shipping (ICS), 2020. Shipping and World Trade. http://www.ics-shipping.org/shipping-facts/shipping- and-world-trade.

15. Heritage Daily, 2017. 10 Ancient Sunken Cities and Settlements, 9 October; https://www.heritagedaily.com/2017/10/10-ancient- sunken-cities-settlements/116944.

16. Hazell, W., 2019. 10 Sunken Cities (that Are Not Atlantis). 8 June, *Listverse*. https://listverse.com/2019/02/18/10-unken-cities-that- are-not-atlantis/.

17. UN, 1982. United Nations Convention on the Law of the Sea, p. 2020. https://www.un.org/depts/los/convention_agreements/texts/ unclos/unclos_e.pdf.

18. Wikipedia, 2020a. United Nations Convention on the Law of the Sea, updated 23 May 2020. https://en.wikipedia.org/wiki/ United_Nations_Convention_on_the_Law_of_the_Sea.

19. UN, 2019. https://www.un.org/Depts/los/reference_files/UNCLOS %20Status%20table_ENG.pdf.

20. Boyd, A. Ptolemy's Geographia, *Engines of Ingenuity*, University of Houston, accessed 23 March 2020. https://uh.edu/engines/epi2594. htm.

21. Israel, R., 2003. Mercator's Projection. https://www.math.ubc.ca/ ~israel/m103/mercator/mercator.html.

22. Wikipedia, 2020b. Depth Sounding. https://en.wikipedia.org/wiki/ Depth_sounding.

23. Government of India, Ministry of Earth Sciences, 2015. History of Meteorological Services in India. https://web.archive.org/web/20160219093330/http://www.imd.gov.in/pages/about_history.php.

24. Timeline of Meteorology. https://en.wikipedia.org/wiki/Timeline_of_meteorology#cite_note-IMD_History-1. Retrieved 18 March 2020.

25. Frisinger, H. H., 1972. Aristotle and his "Meteorologica'. *Bull. Am. Meteorol. Soc.*, vol. 53, no. 7, pp. 634–638.

26. Graham, S., Parkinson, P., and Chahine, M., 2002. Weather Forecasting through the Ages. 25 February. https://earthobservatory.nasa.gov/features/WxForecasting/wx2.php.

27. Cahir, J. J., 2019. Weather Forecasting. *Encyclopedia Britannica*. 19 January. https://www.britannica.com/science/weather-forecasting.

28. Forbes, E., 1844. Report on the Mollusca and Radiata of the Aegean Sea, and on their distribution, considered as bearing on geology. *Report of the British Association for the Advancement of Science,* for 1843, pp. 129–193, available at: https://www.biodiversitylibrary.org/item/46634#page/176/mode/1up.

29. Rozwadowski, H. M., 2005. *Fathoming the Ocean: The Discovery and Exploration of the Deep Sea*. Chapter 5, Dredging the Moon, pp. 133–174 Harvard University Press. ISBN 978-0-674-01691-0.

30. Burns, B., 2012. Submarine Cable System History. http://www.submarinecablesystems.com/default.asp.pg-history.

31. Golembiewski, K., 2019. HMS Challenger —Humanity's First Real Glimpse of the Deep Oceans. *Discover Magazine*, Planet Earth 19 April. https://www.discovermagazine.com/planet-earth/hms-challenger-humanitys-first-real-glimpse-of-the-deep-oceans.

32. Cline, E., 2014. *1171 B.C. The Year Civilization Collapsed*. Princeton, Princeton University Press, ISBN 9780691140896.

33. Cartwright, M., 2019. The Seven Voyages of Zheng He. Ancient™ History Encyclopedia, 7 February; https://www.ancient.eu/article/1334/the-seven-voyages-of-zheng-he/.

34. Banes, D. (1988). The Portuguese Voyages of Discovery and the Emergence of Modern Science. *J. Washington Acad. Sci.*, vol. 78, no. 1, pp. 47–58. https://www.jstor.org/stable/24536958.

35. NOAA, 2020b. What Are the Trade Winds? https://oceanservice.noaa.gov/facts/tradewinds.html.

36. NOAA, 2020c. What Are the Doldrums? https://oceanservice.noaa.gov/facts/tradewinds.html.

37. Royal Museums Greenwich, 2012. 18th Century Sailing Times between the English Channel and the Coast of America: How Long Did It Take? 8 November. https://www.rmg.co.uk/discover/behind-the-scenes/blog/18th-century-sailing-times-between-english-channel-and-coast-america.

38. Richardson, P. L., and Adams, N. T., 2018. Nantucket Whalers and the Franklin-Folger Chart of the Gulf Stream, *Uncharted Waters*, Nantucket Historical Association, Spring, pp. 17–24. https://www2.whoi.edu/staff/prichardson/wp-content/uploads/sites/75/2018/11/Richardson-Adams-2018-HN-Uncharted-Waters.pdf.

39. Wilkinson, J., 2010. History of the Gulf Stream. *Keys Historeum*. Historical Preservation Society of the Upper Keys. Retrieved 19 May 2020; http://www.keyshistory.org/gulfstream.html.

40. Richardson, P. L., 1982. Walter Hoxton's 1735 description of the Gulf Stream. *J. Mar. Res.*, supplement vol. 40, pp. 597–603. Retrieved from https://images.peabody.yale.edu/publications/jmr/jmr40-S-30.pdf.

41. Op.Cit. reference 28.

42. Wikipedia, 2020c. Benjamin Franklin. Updated to 19 May. https://en.wikipedia.org/wiki/Benjamin_Franklin.

43. Toomey, M., 2018. John William Gerard De Brahm. Tennessee Historical Society. http://tennesseeencyclopedia.net/entries/john-william-gerard-de-brahm/.

44. Maury, M. F., 1855. *The Physical Geography of the Sea*. Sampson, Low, Son & Co. London, UK. Retrieved from https://doi.org/10.5962/bhl.title.102148.

45. *Captain Maury, M. F. Nature,* vol. 7, pp. 390–391 (1873). https://doi.org/10.1038/007390b0.

46. Duncan, S.-K., et al., 2020. Ocean Surface Currents. *Exploring Our Fluid Earth.* https://manoa.hawaii.edu/exploringourfluidearth/physical/atmospheric-effects/ocean-surface-currents.

47. Hadley, G., 1735. Concerning the Cause of the General Trade Winds. *Phil. Trans. Roy. Soc.*, vol. 29, pp. 58–62. https://royalsocietypublishing.org/doi/10.1098/rstl.1735.0014.

48. Ferrel, W., 1858. The influence of the Earth's rotation upon the relative motion of bodies near its surface. *Astron. J.*, 20 January, vol. 5, no. 109, pp. 97–100. http://adsabs.harvard.edu/full/1858AJ......5...97F.

49. Ekman, V. W., 1905. On the Influence of the Earth's Rotation on Ocean Currents. *Arkiv för matematik, astronomi och fysik.* Bd 2, #11, pp. 1–51. https://jscholarship.library.jhu.edu/handle/1774.2/33989?show=full.

50. Wikipedia, 2020d. Fridtj of Nansen. Retrieved 31 May 2020. https://en.wikipedia.org/wiki/Fridtjof_Nansen.

51. Coriolis, G, G., 1835. Mémoire sur les équations du mouvement relatif des systèmes de corps (On the equations of relative motion of a system of bodies). *J. Ec. Polytech.*, vol. 15, pp. 142–154. https://repository.ou.edu/uuid/8cb6fb18-779d-5968-9ddf-80ee7887aad4#page/1/mode/2up.

52. University of Hawai'i, 2020. Ocean Surface Currents. *Exploring Our Fluid Earth*; https://manoa.hawaii.edu/exploringourfluidearth/physical/atmospheric-effects/ocean-surface-currents.

53. National Research Council, 2000. *50 Years of Ocean Discovery: National Science Foundation 1950–2000*. Washington DC: The National Academies Press, p. 283; https://doi.org/10.17226/9702.

54. Mills, E. L., 2009. *The Fluid Envelope of Our Planet: How the Study of Ocean Currents Became a Science*. University of Toronto Press. ISBN 978080209675; e-book ISBN 97814426974. https://books.scholarsportal.info/en/read?id=/ebooks/ebooks2/utpress/2013-08-26/1/9781442697744.

55. Deacon, M., 1971. *Scientists and the Sea, 1650–1900: A Study of Marine Science* (1st ed.), 1971, Academic Press Ltd., London, 2nd edition published 2016 Routledge, London & New York, e-book, 17 April 2018 Routledge, p. 504; ISBN1351901575, 9781351901574.

56. Docevski, B., 2017. Depth Sounding Techniques that Preceded the Modern Day SONAR Technology. *the Vintage News*, 23 February. https://www.thevintagenews.com/2017/02/23/depth-sounding-techniques-that-preceded-the-modern-day-sonar-technology/Day, A., 1967. *The Admiralty Hydrographic Service 1795–1919*, Vice-Admiral Sir Archibald Day, H.M.S.O., London, UK: ASIN: B0000CO478.

57. Day, A., 1967. *The Admiralty Hydrographic Service 1795–1919*, Vice-Admiral Sir Archibald Day, H.M.S.O., London, UK: ASIN: B0000CO478.

58. David, A., 2008. The Emergence of the Admiralty Chart in the Nineteen Century. ICA Commission on the History of Cartography, Symposium on *"Shifting Boundaries: Cartography in the 19th and 20th centuries"* Portsmouth University, Portsmouth, United Kingdom, 10–12 September, p. 16.

59. Scripps Institute of Oceanography UC San Diego, 2020. *About Scripps.* https://scripps.ucsd.edu/about.

60. Marine Biological Association of the UK, 2020. Last accessed 3 June 2020. https://www.mba.ac.uk/about.

61. Oceanic Image Consultants, 2013. Brief History of Sonar Development. p. 4; http://www.oicinc.com/brief-history-sonar.html.

62. Wetzel, R. G., 2001. *Limnology, Lake and River Ecosystems* (3rd ed.), p. 1006, Academic Press, ISBN-13: 978-0-12-744760-5; https://doi.org/10.1016/C2009-0-02112-6.

63. Birge, E. A., 1904. The Thermocline and Its Biological Significance. *Trans. Am. Microsc. Soc.*, September, vol. 25, pp. 5–33, DOI: 10.2307/3220866; https://www.jstor.org/stable/3220866.

64. Simony, F., 1850. *Die Seen des Salzkammergutes*, 31 December, p. 24, Google e-Book, in German: available through https://books.google.com/.

65. Office of Naval Research, 1953. *Notes from the Conference of the Thermocline 25–27 May,* Pollack, M. J., ed. https://apps.dtic.mil/dtic/tr/fulltext/u2/028254.pdf.

66. WHOI, 2020. *Learn about WHOI.* https://www.whoi.edu/who-we-are/.

67. Seiwell, H. R., 1947. Military Oceanography in World War II. *Mil. Eng.*, May, vol. 39, no. 259, pp. 202–210. https://www.jstor.org/stable/44555900.

68. Tunnicliffe, M., 2010. Ocean Acoustics in World War II: Dawn of a New Science in Canada. *J. Ocean Technol.*, Canadian Navy: 100 years, vol. 5, Special Issue, p. 12. https://www.thejot.net/article-preview/?show_article_preview=177.

69. Plymouth Marine Laboratory, 2020. Retrieved 3 June 2020. https://www.pml.ac.uk/.

70. Talley, L. D., et al., 2011. *Descriptive Physical Oceanography: An Introduction.* 6th ed., p. 560, Elsevier-Academic Press, ISBN 978-0-7506-4552-2; https://www.sciencedirect.com/book/9780750645522/descriptive-physical-oceanography.

71. NOAA, 2019a. What Is a Thermocline? https://oceanservice.noaa.gov/facts/thermocline.html.

72. NOAA, 2020d. NOAA-NESDIS Geo-Blended 5 km SST Analysis, Contoured Image for 23 June 2020. *Office of Satellite and Product Operations.* https://www.ospo.noaa.gov/data/sst/contour/global_small.cf.gif.

73. OpenLearn, 2016. The Oceans. Open University (UK), p. 50. https://www.open.edu/openlearn/science-maths-technology/the-oceans/content-section-0?active-tab=description-tab.

74. Sverdrup, H. U., Johnson, M. W., and Fleming, R. H., 1942. *The Oceans, Their Physics, Chemistry and General Biology.* Prentice-Hall. https://wwnorton.com/college/geo/oceansci/ch/06/welcome.asp.

75. NOAA, 2020e. What Is Forchhammer's Principle? https://oceanservice.noaa.gov/facts/forchhammers-principle.html.

76. Op.Cit [70], 2011, Chapter 3, p 35.

77. Shkvorets, I., 2020. Early Determination of Salinity: From Ancient Concepts to Challenger Results. *Salinometry.* https://salinometry.com/early-determination-of-salinity-from-ancient-concepts-to-challenger-results/.

78. Bjornerud, M., 2019. Reading Seawater. *Inference: Int. Rev. Sci.,* 13 December, vol. 5, no. 1; https://inference-review.com/article/reading-seawater.

79. NOAA, 2020f. Why Is the Ocean Salty? Last updated 11 May. https://oceanservice.noaa.gov/facts/whysalty.html.

80. Denchak, M., 2018. Water Pollution: Everything you need to know. *National Resources Defense Council: Our Stories,* 14 May. https://www.nrdc.org/stories/water-pollution-everything-you-need-know.

81. Segar, D. A., 2007. *Introduction to Ocean Sciences* (2nd ed.), p. 720; ISBN-10: 039392629X. https://wwnorton.com/college/geo/oceansci/.

82. CATDS, 2020. Centre Aval de Traitement des Données SMOS, Sea Surface Salinity Measurements: Definitions and Units. http://www.salinityremotesensing.ifremer.fr/sea-surface-salinity/definition-and-units. Retrieved 27 June 2020.

83. Research Gate, 2020; retrieved 27 June https://www.researchgate.net/post/Does_anyone_have_experience_with_practical_salinity_scale_and_Practical_Salinity_Unit.

84. NASA, 2020b. Ocean Motion: Ocean's Vertical Structure. http://oceanmotion.org/html/background/ocean-vertical-structure.htm.

85. Roquet, F., et al., 2015. Defining a Simplified but "Realistic" Equation of State of Seawater. *J. Phys. Oceanogr.,* vol. 45, no. 10, pp. 2564–2579. https://doi.org/10.1175/JPO-D-15-0080.1.

86. Millero, F. J., 2010. History of the Equation of State of Seawater. *Oceanography,* vol. 23, no. 3, pp. 18–33, doi:10.5670/oceanog.2010.21.

87. Cenedese, C., and Gordon, A. L., 2018. Ocean Current. *Encyclopedia Britannica,* 30 May. https://www.britannica.com/science/ocean-current.

88. Rahmstorf, S., 2013. Glacial Climates: Thermohaline Circulation, in *Encyclopedia of Quaternary Science* (2nd ed.), Elias, S. A., and Mock, C. J., eds., pp. 737–747, Elsevier, ISBN-13:978-0444536433.

89. National Geographic, 2020. Ocean Gyre. Resource Library: Encyclopedic Entry, Last Retrieved 2 July. https://www.nationalgeographic.org/encyclopedia/ocean-gyre/.

90. Wikipedia, 2020e. Ocean Gyre. https://en.wikipedia.org/wiki/Ocean_gyre; last edited 1 June, diagram by Avsa—Own work, CC BY-SA 3.0. https://commons.wikimedia.org/w/index.php?curid=8385258.

91. NASA, 2020c. Ocean Motion: Wind Driven Surface Currents: Gyres. last retrieved 4 July. http://oceanmotion.org/html/background/wind-driven-surface.htm.

92. UN, 2016. Sea-Air Interactions, in *The First Global Integrated Marine Assessment World Ocean Assessment I* (Innis, L, and Simcox, A., et al.), Chapter 5. https://www.un.org/Depts/los/global_reporting/WOA_RPROC/WOACompilation.pdf.

93. NIWA, 2020. Layers of the Atmosphere. *NIWA Taihoro Nukurangi.* Retrieved 6 July. https://niwa.co.nz/education-and-training/schools/students/layers.

94. Wikipedia, 2020f. Kármán line. Updated to 25 May. https://en.wikipedia.org/wiki/K%C3%A1rm%C3%A1n_line.

95. Riebeek, H., and Simmon, R., 2009. Catalog of Earth Satellite Orbits, 4 September, NASA earth observatory. https://earthobservatory.nasa.gov/features/OrbitsCatalog.

96. Scott, A., 2016. The What-O-Sphere? An Explainer. *The Planetary Society*, 5 May. https://www.planetary.org/blogs/guest-blogs/2016/0505-the-what-o-sphere-an-explainer.html.

97. Diallo, M., et al., 2017. Global Distribution of CO_2 in the Upper Troposphere and Stratosphere. *Atmos. Chem. Phys.*, vol. 17, pp. 3861–3878, 2017. https://doi.org/10.5194/acp-17-3861-2017.

98. Salawich, R. J., et al., 2019. Twenty Questions and Answers About the Ozone Layer: 2018 Update, Scientific Assessment of Ozone Depletion: 2018, 84 p., World Meteorological Organization, Geneva, Switzerland, 2019. Published in November 2019 ISBN: 978-1-7329317-2-5.

99. CCAR, 2020. Tropospheric Ozone. Retrieved 10 July 2020; https://www.ccacoalition.org/en/slcps/tropospheric-ozone.

100. Halley, E., 1687. An Historical Account of the Trade Winds, the Monsoons, Observable in the Seas between and Near the Tropicks,

with an Attempt to Assign the Physical Cause of the Said Winds. *R. Soc. Philos. Trans.*, vol. 16, pp. 153–169. https://doi.org/10.1098/rstl.1686.0026.

101. The Editors of the Encyclopaedia Britannica, 2016. Hadley Cell. 20 January Retrieved 12 July 2020. https://www.britannica.com/science/Hadley-cell.

102. NOAA, 2020g. What Are the Roaring Forties? Last Update 23 April; https://oceanservice.noaa.gov/facts/roaring-forties.html.

103. Persson, A. O., 2006. Hadley's Principle: Understanding and Misunderstanding the Trade Winds. *Hist. Meteorol.*, vol. 3, pp. 17–42. http://www.meteohistory.org/2006historyofmeteorology3/2-persson_hadley.pdf.

104. Rochas, M., 2020. Teisserenc de Bort, Léon Philippe. https://www.encyclopedia.com/science/dictionaries-thesauruses-pictures-and-press-releases/teisserenc-de-bort-leon-philippe.

105. Cave, C., 1928. Mr. W. H. Dines, FRS. *Nature,* 14 January, vol. 121, pp. 65–66; https://doi.org/10.1038/121065a0.

106. Fueglistaler, S., et al., 2009. Tropical Tropopause Layer. *Rev. Geophys.*, vol. 47, RG1004, doi:10.1029/2008RG000267; https://agupubs.onlinelibrary.wiley.com/doi/pdf/10.1029/2008RG000267.

107. Piana, M. E., 2020. Hadley Cells. Last retrieved 10 July. https://www.seas.harvard.edu/climate/eli/research/equable/hadley.html.

108. Moreau, R., 2017. Air and Water: *Trade Winds, Hurricanes, Gulf Stream, Tsunamis and Other Striking Phenomena.* Springer International (1st ed.), ISBN-10: 3319652133.

109. Lumen, 2020. Physical Geography—Global Atmospheric Circulations. Retrieved 16 July. https://courses.lumenlearning.com/geophysical/chapter/global-atmospheric-circulations/.

110. NOAA, 2008. Cross section of the two main jet streams, by latitude. *National Weather Service*, 8 May. http://www.srh.noaa.gov/jetstream//global/images/jetstream3.jpg.

111. Wikipedia, 2020g. Köppen Climate Classification. Retrieved 15 July, p. 27. https://en.wikipedia.org/wiki/K%C3%B6 Open Climate Classification.

112. Belda, M., et al., 2014. Climate Classification Revisited: From Köppen to Trewartha. *Clim. Res.*, vol. 59, pp. 1–13, 2014 doi: 10.3354/cr01204; https://www.int-res.com/articles/cr_oa/c059p001.pdf.

113. Arnfield, J., 2020. Köppen Climate Classification. *Encyclopædia Britannica*, 20 January. https://www.britannica.com/science/Koppen-climate-classification.

114. Cain, F., 2016. What is the Largest Desert on Earth? *Universe Today*, 1 September. https://www.universetoday.com/27064/what-is-the-largest-desert-on-earth/.

115. United States Geological Survey (USGS), 2001. What Is a Desert? https://pubs.usgs.gov/gip/deserts/what/.

116. Murigi, E., 2018. What Are The Characteristics Of An Oceanic Type Of Climate? WorldAtlas, 5 February. https://www.worldatlas.com/articles/what-are-the-characteristics-of-an-oceanic-type-of-climate.html.

117. The World Factbook, 2020. *Central Intelligence Agency*, Washington, DC. https://www.cia.gov/library/publications/the-world-factbook/fields/286.html.

118. Stull, R., 2017. Practical Meteorology: An Algebra-Based Survey of Atmospheric Science. Version 1.02b. p. 944. ISBN 13: 978-0-88865-283-6; https://www.eoas.ubc.ca/books/Practical_Meteorology/.

119. Rankin, B., 2016. World Population Distribution by Altitude. *Radical Cartography*. http://www.radicalcartography.net/howhigh.html.

120. Lykossov, V. N., 2001. Atmospheric and Ocean Boundary Layer Physics, in *Wind Stress Over the Ocean*, Jones, I. S. F., and Toba, Y., eds., Book chapter, pp. 54–81, Cambridge University Press. https://doi.org/10.1017/CBO9780511552076.004.

121. Broecker, W. S., 1991. The Great Ocean Conveyor. *Oceanography*, vol. 4, no. 2, pp. 79–89. https://tos.org/oceanography/assets/docs/4-2_broecker.pdf.

122. Stommel, H., 1948. The westward intensification of wind-driven ocean currents. *Eos Trans. Am. Geophysical Union*, vol. 29, no. 2, pp. 202–206. https://doi.org/10.1029/TR029i002p00202.

123. National Academy of Sciences, 1997. Henry Stommel. *Biographical Memoirs*, vol. 72, Washington DC, National Academies Press, doi 10.17226/5859.

124. NOAA, 2020h. The Younger Dryas. Last accessed 21 July. https://www.ncdc.noaa.gov/abrupt-climate-change/The%20Younger%20Dryas.

125. Broecker, W., 2010. The Great Ocean Conveyor: *Discovering the Trigger for Abrupt Climate Change*. Princeton University Press, pp. 176, ISBN 9780691123545. https://press.princeton.edu/books/hardcover/9780691143545/the-great-ocean-conveyor.

126. Webb, P., 2020. Introduction to Oceanography. Updated July, p. 393; Roger Williams University. https://rwu.pressbooks.pub/webboceanography/chapter/9-5-currents-upwelling-and-downwelling/.

127. Kantha, L., and Clayson, C. A., 2003. Ocean Mixed Layer, in *Encyclopedia of Atmospheric Sciences*, ed., by Holton, J. R., Curry, J. A., and Pyle, J. A., pp. 291–298; Academic Press, ISBN 978-0-12-227090-1.

128. De Boyer Montégut, C., 2015. Mixed Layer Depth Climatology and Other Related Ocean Variables. *Ifremer*; http://www.ifremer.fr/cerweb/deboyer/mld/Surface_Mixed_Layer_Depth.php.

129. Kloster, S., et al., 2006. DMS Cycle in the Marine Ocean-Atmosphere System—A Global Model Study. *Biogeosciences*, vol. 3, no. 1, pp. 29–51. https://doi.org/10.5194/bg-3-29-2006.

130. Pellichero, V., et al., 2017. The Ocean Mixed Layer under Southern Ocean Sea-Ice: Seasonal Cycle and Forcing. *J. Geophys. Res. Oceans.*, vol. 122, pp. 1608–1633; doi:10.1002/2016JC011970.

131. WWF, 2020. Water Scarcity. https://www.worldwildlife.org/threats/water-scarcity.

132. Gruère, G., Shigemitsu, M., and Crawford, S., 2020. Agriculture and Water Policy Changes: Stocktaking and Alignment with OECD and G20 Recommendations, OECD Food, Agriculture and Fisheries Papers, No. 144, *OECD Publishing*, Paris. http://dx.doi.org/10.1787/f35e64af-en.

133. Nagaraja, M. P., 2020. Water Cycle. NASA Science. https://science.nasa.gov/earth-science/oceanography/ocean-earth-system/ocean-water-cycle.

134. Northwest River Forecast Centre, 2020. Description of the Hydrologic Cycle. Retrieved 21 July 2020. https://www.nwrfc.noaa.gov/info/water_cycle/hydrology.cgi.

135. Pope, E. C., Bird, D. K, and Rosing, M. T., 2012. Isotope composition and volume of Earth's early oceans. *PNAS* ⬚ March 20, vol. 109, no. 12, pp. 4371–4376; https://doi.org/10.1073/pnas.1115705109.

136. Lagerloef, G., et al., 2010. The Ocean and the Global Water Cycle. *Oceanography,* vol. 23, no. 4, Special Issue on the Future of Oceanography From Space, pp. 82–93; https://www.jstor.org/stable/24860864.

137. UKMetOffice, 2018. What is Global Circulation, Part Two, The Three Cells. 20 February. https://www.youtube.com/watch?v=xqM83_og1Fc&t=53s.

138. Wikipedia, 2020h. The Climate of Europe. Updated 26 May. https://en.wikipedia.org/wiki/Climate_of_Europe.

139. Li, Z., et al., 2016. Aerosol and monsoon climate interactions over Asia. *Rev. Geophys.*, vol. 54, pp. 866–929, doi: 101002/2015RG00050.

140. North Carolina Climate Office, 2020. Monsoon Systems. Retrieved 27 July. https://climate.ncsu.edu/edu/Monsoons.

141. Osborn, L., 2020. United States' Rainiest Cities. *Current Results*. https://www.currentresults.com/Weather-Extremes/US/wettest-cities.php.

142. Saha, K., 2010. *Tropical Circulation Systems and Monsoons*. Springer-Verlag, ISBN 978-3-642-03372-8, doi 10.1007/978-3-642-03372-5.

143. WRCP, 2020. The Global Monsoon Systems. *The World Climate Research Programme*. https://www.wcrp-climate.org/documents/monsoon_factsheet.pdf.

144. WMO, 2005. *The Global Monsoon System: Research and Forecast Report*, WMO/TD No. 1266 (TMRP Report No. 70), Chang, C. P., Wang, B., and Lau, N. C. G., eds., p. 552. https://www.wmo.int/pages/prog/arep/tmrp/documents/global_monsoon_system_IWM3.pdf.

145. USGS, 2020. West Africa: Land Use and Land Cover Dynamics. Retrieved 30 July. https://eros.usgs.gov/westafrica/population.

146. Raj, J., Bangalath, H. K., and Stenchikov, G., 2019. West African Monsoon: Current State and Future Projections in a High-Resolution AGCM. *Clim. Dyn.*, vol. 52, pp. 6441–6461. https://doi.org/10.1007/s00382-018-4522-7.

147. Landsea, C. W., and Gray, W. M., 1992. The Strong Association between Western Sahelian Monsoon Rainfall and Intense Atlantic Hurricanes. *J. Clim.*, vol. 5, no. 5, pp. 435–453. https://doi.org/10.1175/1520-0442(1992)005<0435: TSABWS>2.0.CO;2.

148. NOAA, 2020i. What is Storm Surge? Last updated 20 May; https://oceanservice.noaa.gov/facts/stormsurge-stormtide.html.

149. Ghose, T., 2019. Hurricane Season: How Long It Lasts and What to Expect. *Live Science*, 23 May. https://www.livescience.com/57671-hurricane-season.html.

150. WHOI, 2020a.El Niño and other oscillations. https://www.whoi.edu/know-your-ocean/ocean-topics/ocean-circulation/el-nio-other-oscillations/.

151. Pacific Marine Environmental Laboratory, 2020. El Niño Theme Page—Impacts. Retrieved 30 July 2020. https://www.pmel.noaa.gov/elnino/impacts-of-el-nino.

152. Katz, R. W., 2002. Sir Gilbert Walker and a Connection between El Niño and Statistics. *Stat. Sci.*, vol. 17, no. 1, pp. 97–112. https://projecteuclid.org/download/pdf_1/euclid.ss/1023799000.

153. L'Heureux, M., 2014. What is the El Niño–Southern Oscillation (ENSO) in a Nutshell? https://www.climate.gov/news-features/blogs/enso/what-el-ni%C3%B1o%E2%80%93southern-oscillation-enso-nutshell.

154. Australian Bureau of Meteorology, 2020. El Niño Southern Oscillation (ENSO). http://www.bom.gov.au/climate/about/index.shtml?bookmark=enso.

155. Di Leberto, T., 2014. The Walker Circulation: ENSO's atmospheric buddy. https://www.climate.gov/news-features/blogs/enso/walker-circulation-ensos-atmospheric-buddy.

156. WMO, 2020. Weather Forecasts and Warnings. https://public.wmo.int/en/our-mandate/weather/weather-forecast.

157. Stull, R., 2017. *Op. Cit.,* reference 118, p825.

158. McPhee, M., 2008. Air-Ice-Ocean Interaction. *Springer*, p. 192, ISBN 978-0-387-78335-2.

159. Scott, M., and Hansen, K., 2016. Sea Ice. *NASA Earth Observatory*, 16 September. https://earthobservatory.nasa.gov/features/SeaIce.

160. UN, 1992. United Nations Framework Convention On Climate Change. p. 25, Article 1.2. https://unfccc.int/resource/docs/convkp/conveng.pdf.

161. NOAA, 2020j. Defining Climate Normals in New Ways. National Centers for Environmental Information. Retrieved 20 July. https://www.ncdc.noaa.gov/news/defining-climate-normals-new-ways.

162. USEPA, 2016. Climate Change Indicators: Oceans, 2 August. https://www.epa.gov/climate-indicators/oceans.

163. Gagosian, R. B., 2003. Abrupt Climate Change: Should We Be Worried? p. 8, report prepared for the World Economic Forum, Davos 27 January 2003. https://www.whoi.edu/wp-content/uploads/2019/04/Abruptclimatechange_7229.pdf.

164. National Research Council, 2002. Abrupt Climate Changes: Inevitable Surprises. *National Academies Press*, Washington DC, p. 252, ISBN 978-0-309-07434-6, DOI 10.17226/10136. http//nap.edu/10136.

165. Clark, P. U., et al., 2002. The Role of the Thermohaline Circulation in Abrupt Climate Change. *Nature*, vol. 425, 21 February, pp. 863–869. https://www.nature.com/articles/415863a.

166. Anderson, D. G., et al., 2011. Multiple Lines of Evidence for Possible Human Population Decline/Settlement Reorganization during the Early Younger Dryas. *Quaternary Int.*, vol. 242, no. 2, 15 October, pp. 570–583; https://doi.org/10.1016/j.quaint.2011.04.020.

167. Watts, A., 2010. Atlantic Conveyor Belt—Still Going Strong and Will Be the Day After Tomorrow. *Watts Up With That?* 29 March. https://wattsupwiththat.com/2010/03/29/atlantic-conveyor-belt-still-going-strong-and-will-be-the-day-after-tomorrow/.

168. Brauer, A., et al., 2008. An Abrupt Wind Shift in Western Europe at the Onset of the Younger Dryas Cold Period. *Nat. Geosci.,* vol. 1, pp. 520–523. https://doi.org/10.1038/ngeo263.

169. Rowe, P. J., et al., 2020. Multi-Proxy Speleothem Record of Climate Instability During the Early Last Interglacial in Southern Turkey. *Palaeogr. Palaeoclimatol. Palaeoecol.,* 15 January, vol. 538. https://doi.org/10.1016/j.palaeo.2019.109422.

170. United Nations, 2017. Factsheet: People and Oceans. https://www.un.org/sustainabledevelopment/wp-content/uploads/2017/05/Ocean-fact-sheet-package.pdf.

171. CoastAdapt, 2017. What Are the RCPs? https://coastadapt.com.au/sites/default/files/infographics/15-117-NCCARFINFOGRAPHICS-01-UPLOADED-WEB%2827Feb%29.pdf.

172. Schmitz, A., and Hunt, B., 2019. https://www.businessinsider.com/what-if-all-ice-earth-melted-overnight-2019-10.

173. Meyer, R., 2019. A Terrifying Sea-Level Prediction Now Looks Far Less Likely. *The Atlantic.* https://www.theatlantic.com/science/archive/2019/01/sea-level-rise-may-not-become-catastrophic-until-after-2100/579478/.

174. NOAA, 2020k, Gravity: Global Positioning Tutorial. https://oceanservice.noaa.gov/education/tutorial_geodesy/geo07_gravity.html.

175. MIT, 2014. Earth's Gravity Field and Sea Level. https://ocw.mit.edu/courses/earth-atmospheric-and-planetary-sciences/12-808-introduction-to-observational-physical-oceanography-fall-2004/lecture-notes/course_notes_3b.pdf.

176. Lemoine, F. G., et al., 1998. The Development of the Joint NASA GSFC and the National Imagery and Mapping Agency (NIMA) Geopotential Model EGM96, p. 584. https://ntrs.nasa.gov/archive/nasa/casi.ntrs.nasa.gov/19980218814.pdf.

177. Pavlis, N. K., et al., 2012. The Development and Evaluation of the Earth Gravitational Model 2008 (EGM2008). *J. Geophys. Res. Solid Earth,* vol. 117, no. B4, April 2012. https://doi.org/10.1029/2011JB008916.

178. Fraczek, W., 2003. Mean Sea Level, GPS, and the Geoid. *Esri ArcUser,* Special Section; July-September issue. https://www.esri.com/news/arcuser/0703/geoid1of3.html.

179. Gregory, J. M., et al., 2019. Concepts and Terminology for Sea Level: Mean. Variability and Change, Both Local and Global. *Surv. Geophys.*, vol. 40, pp. 1251–1289. https://doi.org/10.1007/s10712-019-09525-z.

180. Hamlington, B, Thompson, P., et al., 2016. The Climate Data Guide: Tidal Gauge Sea-Level Data. National Center for Atmospheric Research (NCAR/UCAR). https://climatedataguide.ucar.edu/climate-data/tide-gauge-sea-level-data.

181. PSMSL, 2020. Permanent Service for Mean Sea Level. https://www.psmsl.org/.

182. NOAA-Tides & Currents, 2020. Sea Level Trends. https://tidesandcurrents.noaa.gov/sltrends/.

183. Lindsey, R., 2019. Climate Change: Global Sea Level. https://www.climate.gov/news-features/understanding-climate/climate-change-global-sea-level.

184. Ocean and Science Platform, 2020. The Ocean, A Carbon Sink. https://ocean-climate.org/?page_id=3896&lang=en.

185. Jiang, L.-Q., et al., 2019. Surface ocean pH and buffer capacity: past, present and future. *Sci Rep.*, vol. 9, p. 18624. https://doi.org/10.1038/s41598-019-55039-4.

186. Donet, S. C., et al., 2009. Ocean Acidification: The Other CO_2 Problem. *Ann. Rev. Mar. Sci.*, vol. 1, pp. 169–192. https://doi.org/10.1146/annurev. marine.010908.163834.

187. USGS, 2020a, pH and Water. Retrieved 6 August 2020. https://www.usgs.gov/special-topic/water-science-school/science/ph-and-water?qt-science_center_objects=2#qt-science_center_objects.

188. NOAA-PMEL, 2020. Quality of pH Measurements in the NODC Data Archives. Retrieved 6 August 2020. https://www.pmel.noaa.gov/CO₂/story/Quality+of+pH+Measurements+in+the+NODC+Data+Archives.

189. Findlay, H., 2015. Ocean Acidification: De-gassing the myths. *Royal Society of Chemistry*, On-Line Lecture. https://www.youtube.com/watch?v=OyZATKk6Z4A&t=616s.

190. Canadian Climate Forum, 2017. Ocean Acidification. *Canadian Climate Forum*, Issue paper #6, pp. 12, Spring. https://climateforum.ca/wp-content/uploads/2017/04/CCF-IP_6-OA-MAR-30-2017-SCREEN-FINAL.pdf.

191. Ritchie, H., 2019. Seafood Production. *Our World in Data*. https://ourworldindata.org/grapher/capture-fisheries-vs-aquaculture-farmed-fish-production.

192. Petruzzello, M., 2019. Eutrophication. 25 March. https://www.britannica.com/science/eutrophication#ref235287.

193. Burton, R., 2018. Blue-Green Algae: Cyanobacteria. Thompson Earth Systems Institute. 15 July. https://www.floridamuseum.ufl.edu/earth-systems/blog/blue-green-algae-cyanobacteria/.

194. University of Minnesota. Eutrophication of Lakes Will Significantly Increase Greenhouse Gas Emissions. ScienceDaily. ScienceDaily, 26 March 2019. https://www.sciencedaily.com/releases/2019/03/190326081426.htm.

195. USEPA, 2016. *Climate Change Indicators in the USA* 4th ed., p. 96. https://www.epa.gov/climate-indicators/downloads-indicators-report.

196. International Union for Conservation of Nature (IUCN), 2020. Ocean Warming. https://www.iucn.org/resources/issues-briefs/ocean-warming.

197. Laffoley, D., and Baxter, J. M., eds., 2016. *Explaining Ocean Warming: Causes, Scale, Effects and Consequences: IUCN*, p. 460. https://portals.iucn.org/library/sites/library/files/documents/2016-046_0.pdf.

198. Alexander, M. A., et al., 2018. Projected Sea Surface Temperatures Over the 21st Century: Changes in the Mean, Variability and Extremes of Large Marine Ecosystem Regions of North Oceans. *Elementia*, 26 January. https://online.ucpress.edu/elementa/article/doi/10.1525/elementa.191/112778/Projected-sea-surface-temperatures-over-the-21st.

Chapter 2

Environmental Impact Assessment of the Roof Insulation Materials during Life Cycle

Lutfu S. Sua[a] and Figen Balo[b]

[a]*Department of Management and Marketing,*
Southern University, Louisiana, USA
[b]*Department of Industrial Engineering,*
Fırat University, Turkey

lutsua@gmail.com, fbalo@firat.edu.tr

Energy is considered as one of the main factors in wealth production and also an important agent in economic improvement. During the past 20 years, the reality and risk of environmental disruption have become more obvious. The growing ecological troubles are owing to a few factors' combination since the ecological consequences of human behavior have dramatically grown due to the sheer rise in global population, industrial activity, consumption, etc. For sustainable development, providing solutions to ecological troubles that we are facing today needs long term planning and actions.

Roof insulating materials are one of the most vigorous equipment for the constructor and the designer to accomplish maximum energy performance in constructions. Nonetheless, there are also negative consequences from the initial steps of

Climate Change and Pragmatic Engineering Mitigation
Edited by Jacqueline A. Stagner and David S.-K. Ting
Copyright © 2022 Jenny Stanford Publishing Pte. Ltd.
ISBN 978-981-4877-97-8 (Hardcover), 978-1-003-25658-8 (eBook)
www.jennystanford.com

their generation till the final of their beneficial lifetime, i.e., when a construction is demolished or rebuilt. Maintainability stays necessary for two reasons: one is the development of the standard of living and the other is the right salesmanship of the products and the systems. For example, with all these claims, counterclaims, and complexity, a roof planner facing a decision to make the choice of insulation material has to consider environmental protection. This is why another important stage of transparency is to have an "Environmental Product Declaration." This is expensive and complex, yet supplies an ISO confirmed method of comparison for all ecological effects of a system or product over its life, from feedstock extraction process to the end-of-life eradication. To establish an equal and fair comparison, end product group rules are founded to standardize the comparison and evaluation. For example, for the roof insulation materials, the measure's standard unit is 1×1 m, with an R value of 1.

In this chapter, a quantitative framework was developed to choose the most environment-friendly roof insulation material for a roof system design. By comparing the full environmental effects of a product, four different roof insulation materials were analyzed based on experts' choices on eight characteristics of these materials. Full environmental effect data is obtained from the roof insulation material producers' Environmental Product Declaration and Health Product Declaration.

2.1 Introduction

Energy-efficiency and eco-design are pressing ideas that convey the need to search for fresh, environment-friendly construction alternatives that contribute to material decrease and energy consumption. Increasing demand for energy and awareness of the environment around the globe made it imperative to use energy sources much more effectively. Insulation makes excellent financial sense because it decreases buildings' power usage. During a building's life cycle, insulation as a sole investment pays for itself many times. Decreased energy expenditure also contributes significantly to the ecology. A building's internal walls and roof are the interface between its exterior and the interior. The heat loss of a construction occurs through its roof, windows/doors, walls, and floor [1].

Insulation of the roof and building envelopes is the most effective tool for cost effectiveness while controlling the outer components in order to make homes more convenient [2].

Innovative energy ideas and passive methodologies for developing the energy performance of constructions have become a strategic, economic, and ecological topic. In fact, the external wall layers of many existing constructions are insufficient to contrast the solar thermal gains in summer months, or to decrease thermal losses in winter months [3, 4]. Especially, the energy cooling requirement and the performance of the roof superficies highly impacts the inside heat comfort in non-conditioned constructions as well as the load in conditioned constructions [5–7]. Furthermore, in summer months the sun irradiation that hits a roof induces the rise of the outside superficies temperature by a few degrees above the outside air temperature [8–10]. The roofs require big quantities of heat gain as they have big surface compared to the many other construction parts. The current placement methodologies contain implying professional insulating board to the roof's surfaces.

Not only does the roof insulation makes it feel more homely, but it also maintains heat throughout the entire house, possibly saving cash on the heating bills and providing a pleasant space throughout the year [11].

Because insulation lowers energy consumption, it offers continuous environmental advantages throughout the life of a building. While insulation products have a beneficial effect on the environment by decreasing power usage in houses, their 'combined energy' also has some adverse environmental effects. It is essential to define building insulation materials that will have minimal adverse effects on the environment [12].

The individual layers must be matched to each other for a correct roof structure. Insulation is mounted between the roof rafters for inter-rafter insulation. Insulation from the inside is mounted below-rafter [13]. This would provide much higher energy protection in the construction fabric. It is suggested to select environment-friendly materials to execute the task set while minimizing all effects, including pollution. Direct comparisons of the different pollution hazards are hard to make, but the manufacturing and application data related to the various roof materials give an indication of the pollutant comparison.

The requirement for developed constructions' energy performance compels the constructors and the designers to utilize materials with low (lower than 0.04 W/mK) k coefficients, which simultaneously require executing maximum performances related to a series of physical and mechanical features. The most often utilized ones are the inorganic fiber insulation materials (stone and glass wool), the organic foams insulation materials (polyurethane and extruded-expanded polystyrene), while other materials encase the residuary 10% of the market (primary wool-wood). Foam-glass and cork are utilized in private status with high needs for private physical features, primarily humidity and pressure strength [13].

Heat insulating is the main tool to improve a construction's energy performance, especially in the heavily built urban areas, where the energy plan's fundamental principles, such as insulation, orientation, and sun-preservation are not readily practical. The insulation materials make constructions more energy-efficient; decrease the fossil-based energy source quantities required for cooling/heating and by that means decrease the amount of CO_2 and SO_2 released into the sky, especially on a meso- and micro scale. Additionally, given the effective urbanization observed within the twentieth century, as CO_2 is one of the paramount greenhouse gases conducing to worldwide warming and SO_2 is the acid rain's main component, the energy management plays an important role on a worldwide ecological scale [14].

The systematic endeavors made toward more power-effective constructions stem from 1970's fuel crises. Heat insulation became compulsory in most countries and the nationwide legal model has become more stringent thenceforward. The involvement was concentrated on the technical features of the materials utilized and the construction's energy attitude for 20 years. In the procedure of applying environmental protection and energy conservation, caution altered to more ecology-friendly construction materials, insulation materials being one of the most attractive fields. The lifecycle investigation strategy concentrates (in that case insulation materials) on the combined power, the re-cycling/

re-use alternatives of the insulation material or its secure eradication.

Based on the materials prevailing the market, the inorganic fiber insulation materials are less energy depleting in their generation and simpler to manipulate as a residual product. The organic foam insulation materials are more power depleting, simpler to manipulate in the structure, yet this combined power can be re-gained if they are reused as main energy resources or raw material. Actually, considering that both groups of materials are evenly efficient in their insulation importance, the last selection relies on the logistics and the economics of the whole loop. The inorganic fiber insulation materials are significantly more economic for the builder, yet do not permit earnings at the destruction. On the other hand, the organic foam insulation materials are less economic, yet permit profit-making repurchase strategies. The transportations of collecting and separating the materials rely heavily on national and local statuses, such as the population density and the urbanization degree. Purpose of an efficient strategy should be to utilize these elements to activate a possible strategy of long-term improvement in the construction industry.

Ekincioglu and co-workers claim that the environment-friendly building structures and effective utilization of restricted sources is very significant. In general, in the EU and the U.S., "maintainability" foundations are noted in the building sector. Several certificates exist to coordinate the characteristics of maintainable structure. Ekincioglu and co-workers state that BREEAM (Building Research Establishment Environmental Assessment Method), BRE (Building Research Establishment) has improved and LEED (Leadership in Energy and Environmental Design), the American system, that is improved by USGBC are two known certification networks. Other certification networks in other countries like Australia, Canada, and Japan are utilized. In Turkey, evaluation and recognition of ecological efficiency of the construction and buildings materials has also been started [15].

For the disposal and collection of insulation materials, there are extra residual administration expenses for the materials which are not being re-cycled after the end of their beneficial lifetime. As an alternative, tax discounts are given for re-using and re-cycling the materials. The comparison between inorganic fibrous and organic foam insulation materials is the most attractive point of view in the insulation materials' ecological perspective, especially as these two classes predominate the market.

On this issue, there are a few examples. For example, Dylewski and Adamczyk have shown that it is lucrative, ecologic, and economical to conduct thermal insulation of construction walls at present energy and thermal insulation material prices with reduced heat transfer coefficient values than those imposed by domestic norms [16]. For global warming, a lifetime ecological effect evaluation protocol for diverse insulating materials suggested from the viewpoint of main energy potential and consumption by Shrestha et al. 2014 [17]. It is stated by Asif and co-workers, that the multidisciplinary process is needed for energy-saving, recycling, and reuse of materials. The research contemplates life-cycle power investigation of constructions as a methodology to develop and examine strategies and methodologies to accomplish decline in main energy utilizations of the constructions as well as managing pollutants [18]. Crawford grouped policies that can diminish the environmental effects of the constructions and buildings when noted in the course of the planned operation [19]. Ramesh and coworkers suggest utilizing sustainable powers in place of petroleum-based resources that can diminish environmental impacts severely [20]. As noted by Kotaji et al., construction sector has significant negative impacts on ecology in two primary aspects: excessive resource extraction and ecological deterioration. These affects raise the need for environmentally friendly construction materials [21]. Wright and Mackey [22] treated homogeneous roofs and walls but then extended their original study [23] to contain composite structures. Balo investigated "Ecological insulating material production evaluation through AHP [24].

The aim of this research is to assess the ecological effects of roof insulation materials on a lifetime basis of "Environmental Product Declaration" and "Health Product Declaration." This chapter uses a list of variables to be regarded in the assessment of the indirect and direct ecological effects of construction products for roof insulating. The specialist opinions were used as the primary scheme for assessing the effects of chosen insulation materials on the environment throughout their life. By using Life-Cycle evaluation methodology in conjunction with an AHP methodology, all effects are evaluated to determine the finest roof insulation material from the ecological effect perspective.

2.2 Life-Cycle Evaluation Methodology

The life-cycle evaluation is a methodology of evaluating the ecological effect linked with the overall phases of the life cycle of a commercial process, service, or product. For example, in the case of a fabricated material, the ecological impact from feedstock processing and extraction of the fabricate, to the use of the product and transportation of it to re-cycling or the final eradication of the materials from which it is made (grave) are assessed [25, 26].

A life-cycle evaluation includes the inventory of the materials and energy requirement for the product, process, or service along the value chain in the sector and computes the corresponding emissions to the ecology [26]. The life-cycle assessment therefore assesses the cumulative ecological effects over the projected life cycle. The purpose is to improve and document the general ecological profile of the material [26].

Life-cycle analysis provides many advantages by offering detailed analysis about the research subject. One of such advantages is the advanced visibility and decision-making scheme. The method also offers significant insight into the environmental effect of a material.

On the other hand, it has certain limitations as well. The method relies on a set of assumptions as a way of simplifying the real-world problem. Thus, it can leave out various factors which can have impacts on the subject matter. Life-cycle analysis also generates results based on various environmental factors.

Generally known procedures for carrying out life-cycle assessments are concluded in the series of ecological administration characteristics of the International Organization for Standardization 14000 (ISO 14040–14044).

2.2.1 Definition, Synonyms, Goals, and Purpose

In the literature on scientific and official reports, the life-cycle evaluation is synonymously referred to as life-cycle analysis [25, 27]. It is occasionally called "cradle to grave analysis." The purpose of the life-cycle evaluation is to provide a comparison of the ecological effects of services and goods through quantitative methods [28]. This research utilizes one of such quantitative methods, known as Analytic Hierarchy Process (AHP).

A life-cycle assessment begins with clearly defined aim and scope of the research. Such a definition thus includes technical details to guide the following:

- The functional unit, which clearly states the subject of the study, quantifies the service provided by the system, supplies a reference to which the outputs and inputs can refer, and supplies a base for the comparison/analysis of alternative services or goods [29].
- The boundaries of the system, showing which processes are covered within the system analysis, including whether the system produces by-products that have to be taken into account through system expansion or assignment [30].
- Information quality necessities that specify what types of information should be included and which constraints (completeness, date range, region or county of study, etc.) apply [31].
- The assignment methodologies utilized to partition an environmental load of a process when multiple products or functions share the same process. The choice of the allocation methodology for by-products can importantly influence the results of a life-cycle assessment [32].
- The impact categories, which can include categories such as human toxicity, smog, global warming, and eutrophication [31].

2.2.2 Inventory

When analyzing the life-cycle inventory (LCI), an inventory of the flows from and to nature is created for a product system. The inventory flows include the supply of water, energy, and raw materials as well as the release into water, air, and land. Output and input information is utilized in order to create the technical system's flow model when developing the inventory. The flow modeling is often represented by a flow chart that includes evaluation of activities in the related supply chain and provides the system boundaries' definition. The required data on output and input to create the modeling need to be gathered for the complete set of activities that are included in the system, including the supply chain.

2.2.3 Impact Assessment

Inventory analysis is followed by a life-cycle effect evaluation. This phase of the life-cycle effect evaluation is conducted to assess the importance of ecological effects related to the life-cycle effect flow results. Classic life-cycle effect evaluations consist of the following phases:

- Determining the characterization models, category indicators, and impact categories
- Sorting the inventory parameters, assigning to certain impact categories
- Characterizing the LCI flows in common units according to one of the LCIA methods, to be summed later on in order to obtain an overall impact category.

In life-cycle assessments, the characterization completes the life-cycle effect evaluation analysis. Although this is the last mandatory level according to ISO14044, various optional phases can be included [32]. Weighting and grouping can be carried out relying on the aim of the life-cycle assessment work. Standardization typically compares the study results of the effect classes with the overall effects in the related region. Such an activity includes sorting and classifying the impact categories. Weighting phase includes weighting the various ecological effects relative to each other. This way, they can then be collected up in order to obtain a single number showing the overall ecological effect. ISO14044 usually suggests against weighting and states that "weighting must not be utilized in life-cycle assessment works utilized in comparative statements that are planned to be accessible to the public" [33]. However, this is frequently ignored, which in return results in comparisons that may reflect a high level of subjectivity due to the weighting.

The effects on the life cycle are usually grouped under a few phases of the improvement, production, disposal, and utilization of a material. Such effects are split into initial effects, usage effects and end-of-life effects. The initial effects are the raw materials' extraction, the production (raw materials' conversion into a material), the product's transport to a market or location, the installation/construction and the start of occupancy or utilization [34]. Impact on usage includes the physical impact of the operation of the product or plant (e.g., water, energy) and any renovation, maintenance or repair work that is necessary to continue using the facility or product conclude processing of waste and demolition or recyclable materials.

2.2.4 Interpretation

Life-cycle interpretation is a systematic method for quantifying, identifying, reviewing, and assessing data from the conclusions of the life-cycle inventory or the life-cycle effect evaluation. This is the stage where the analysis results are summarized through a set of results drawn from the work and recommendations provided for the audience. Interpretation phase includes the following steps according to ISO 14040:

- Based on the LCA, key issues are identified,
- From the perspectives of consistency, completeness, and sensitivity checks, the work is evaluated,
- Constraints of the study are provided, conclusions are drawn, and recommendations are provided [35].

Validating the results of the study and providing them to the stakeholders in an objective manner is one of the primary purposes of the life-cycle interpretation. Interpretation phase is initiated with making sure that the results are accurate and they are consistent with the research goals. The best way of doing this is through the data elements' identification which clearly helps to each of the impact categories.

As suggested through the medium of Curran, the goal of the life-cycle assessments interpretation step is, in particular, to identify the option that has the minimum negative ecological effect on air, sea, and land sources [36].

2.3   Information Analyses

Information validity is a constant concern in life-cycle analysis. If life-cycle assessment results are to be valid, the information utilized in the life-cycle inventory must be valid and accurate, and therefore up to date in terms of validity. When comparing a pair of life-cycle assessments for diverse materials, services, or processes, it is crucial that information of the same standard is available for the pair to be compared. If one of the couples, for example, a material has a much more availability of valid information and accurate, it cannot be truly compared to another one that has a less usability of such information [37].

Regarding the timeliness of the information, it was found that the validity of the information may conflict with the time required for the information collection. New materials and methods of production are constantly being developed as a result of research and development activities taking place all over the world in a fast pace, which makes it both difficult and important to apply and define current information. In certain industries, products can be updated particularly fast, which requires fast, continuous information collection.

As mentioned above, the inventory in the life-cycle assessments typically comprises various phases such as material extraction, processing and production, product use and disposal [25, 26]. If they can be identified, environmental impact can be reduced significantly by making changes on the phases that pollute the environment the most. For instance, the most energy intensive phase in the life-cycle assessment of an aircraft or an automobile product is during its use due to the petroleum expenditure during the material life. An efficient way of increasing fuel efficiency is to through the vehicle weight reduction. Therefore, aircraft and car producers can reduce ecological effect by replacing heavier products with lighter ones (for example, aluminum- or C fiber-reinforced materials, with all costs being the same).

2.3.1   Variants

There are various types of life-cycle assessment types such as cradle-to-grave, cradle-to-gate, gate-to-gate, and well-to-wheel. This study utilizes the cradle-to-grave assessment. This is the complete life-cycle assessment from the raw materials extraction to the phase of usage and disposal. For instance, trees produce paper that is recycled into energy-saving cellulose insulation (fiber paper) and can then be utilized as an energy saving equipment on the ceiling of a house for many years, saving two thousand times the energy consumption of fossil fuels during the manufacturing process. Upon completion of their lifetime, the cellulose fibers are replaced and the old fibers are disposed of, possibly burned. All inputs and outputs are taken into account for the life cycle's all phases.

Future research direction aims to utilize another assessment variant which is gaining more popularity. This variant, which is named as "ecologically sound life-cycle assessment," has much wider boundaries in terms of taking into account a broad range of ecological effect, although same approaches are utilized by a conventional life-cycle assessment. It should provide guidance for people activities of smart management through comprehension of the effects on environmental sources and the surrounding ecosystems. Eco-life-cycle assessment is a method that quantifies the regulation and support of services throughout the life cycle of assets and products. This approach divides services into four primary classes: support, regulation, provision, and cultural services [37].

2.4 Quantitative Analysis

The fourth section contains the quantitative analysis that is used for the purpose of life-cycle assessment of the materials under investigation. A number of impacts of these materials from extraction to disposal are taken into account as the cradle-to-grave assessment requires.

Markets have many distinct kinds of roof insulating materials. There are certain environmental circumstances in every technique of acquiring roof insulation materials. It is not simple to determine the most environment-friendly insulation material. The lifetime effect of insulation materials on the environment can be widely split into two classifications:

(i) Direct effects owing to the insulation materials' combined energy.

(ii) Indirect or environmental effects avoided owing to decreased building operating power expenditure due to additional insulation. Thus, it is essential to define roof insulation materials that will result in minimal adverse effects on the environment throughout the lifetime of the insulation.

Table 2.1 Environmental Product Declaration results for different roof insulation materials

Environmental impact	Unit	Closed cell SPF (roofing)	Foamular XPS	Mineral wool board	Poyiso roof insulation
Global warming potential	$kgCO_2$	34.3	60.8	7.94	2.58
Acidification potential	Mol H+	1.073	1.78	2.75	0.0098
Ozone depletion potential	kgCFC-11	167×10^{-9}	36.300×10^{-9}	118×10^{-9}	94.7×10^{-9}
Smog creation potential	kgO_3	0.267	0.208	0.254	0.18
Eutrophication potential	kgN	1.33×10^{-4}	9.85×10^{-4}	9.4×10^{-4}	14×10^{-4}
Primary energy demand	MJ	136.7	80.7	98.7	53.8
Resource depletion	kg	7.84	—	—	5.06
Waste to disposal	kg	0.011	0.857	5.12	3.7
Waste to energy	kg	—	—	—	0.0003
Water use	kg (l)	1065	37.9	1.27	172

The current research is aimed at finding the most proper roof insulation material for low thermal conductivity and low environmental impact construction applications. For evaluation with this aim, Environmental Product Declaration results for several different roof insulation material options are given in Table 2.1 [38]. The alternative materials in the columns are the most common materials used on the roof insulation. The criteria on the rows are the parameters with adverse effects on the environment throughout the lifetime of the insulation material. The standard unit of measurement is 1 m^2 of insulation with *R* value of 1, for roof insulation materials. *R* value reflects a material's resistance against the flow of heat. Thus, higher

R value s indicates better insulation. The insulation's total environmental effect can be scaled with ease through multiplying the environmental effect by the total R value needed.

Economic, energy and strength values for different roof insulation materials are given in Table 2.2. Here, thermal resistance (R value) is utilized resistance of an insulation material unit to thermal transfer or in other words, capability of insulation material to resist heat. R value only relates to the thermal conductivity and thickness of the material.

Table 2.2 Economic, energy and strength values for different roof insulation materials

	Closed cell SPF (roofing)	Foamular XPS	Mineral wool board	Poyiso roof insulation
R value (inch)	5.8	5.0	4.0	5.7
Cost U-value of 0.3 (Euro/m³)	30–36	20–25	11–15	17–19
Compressive strength (psi)	36	25	12	20

For the purpose of determining relative weights of the eight factors, an analytic hierarchy process is utilized. This approach provides the ability to take into consideration the relative weights of various criteria and decision alternatives. The method has the advantage of evaluating the selected criteria and their sub-criteria on different levels of the decision hierarchy. It studies large scale problems by separating them into smaller steps. On the hand, it is open for human errors during various phases such as criteria selection and pairwise comparison of the alternatives.

The model starts with determining a set of related criteria contributing towards the overall goal of the study. The selection of these criteria along with their relative priority values are determined based on a set of expert opinions.

In the first phase of the method, following pairwise comparison of the factors based on expert opinions, the method calculates the relative weights. Table 2.3 shows the decision matrix that is used towards determining the factor weights.

Table 2.3 Decision matrix

Matrix		global warming	acidification	ozone depletion	smog creation	eutrophication	energy demand	waste to disposal	water demand	normalized principal Eigenvector
		1	2	3	4	5	6	7	8	
global warming	1	1	1/3	1	2	1	1/9	5	1/2	9.33%
acidification	2	3	1	3	5	3	2	1/6	4	18.70%
ozone depletion	3	1	1/3	1	3	2	1/2	2	3	10.38%
smog creation	4	1/2	1/5	1/3	1	2	1/9	4	1/4	6.83%
eutrophication	5	1	1/3	1/2	1/2	1	1/9	2	1/2	5.09%
primary energy demand	6	9	1/2	2	9	9	1	5	2	27.39%
waste to disposal	7	1/5	6	1/2	1/4	1/2	1/5	1	1/3	11.96%
water demand	8	2	1/4	1/3	4	2	1/2	3	1	10.32%

Table 2.4 shows the factor weights that are calculated based on the pairwise comparisons provided in Table 2.3. As a result of the necessary calculations, normalized principal eigenvector is provided in the table.

Table 2.4 Relative weights of the factors [39]

AHP Analytic Hierarchy Process (EVM multiple inputs)

n= 8	Number of criteria (2 to 10)		Scale: 1		AHP 1-9
p= 0	selected Participant (0=consol.)	2	7	Consolidated	

Objective Selection of most environment friendly roof material.

Date | Thresh: 1E-08 | Iterations: 12 | EVM check: 5,3E-09

Table	Criterion	Comment	Weights	+/-
1	Global warming		9.3%	12.5%
2	Acidification		18.7%	10.8%
3	Ozone deplet		10.4%	5.7%
4	Smog creation		6.8%	10.2%
5	Eutrophication		5.1%	4.5%
6	Primary energy demand		27.4%	17.2%
7	Waste to disp		12.0%	25.3%
8	Water demand		10.3%	7.5%

The second phase of the methodology involves selection of the best insulation material based on the pre-determined criteria set and their relative weights. Four types off roof materials and their values for each factor are provided in Table 2.5. The top section of Table 2.5 shows the nominal values of the alternatives in terms of each factor. These characteristic values are readily available in material specifications. This step starts with measuring the impact of each roof material on each factor and ranking them based on their total impact score.

Figure 2.1 depicts the resulting factor weights provided in Table 2.4. It can be observed that the "primary energy demand" has the biggest weight (27.4%) while "eutrophication" is ranked last among the others with a weight of 5.1%.

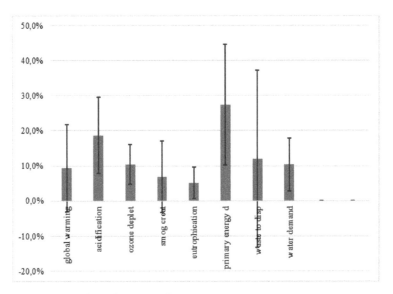

Figure 2.1 Resulting factor weights.

In the mid-section of the table, the values are normalized by dividing each value with the corresponding total value of each row (factor) and these values are then multiplied with the relative weights of each factor which were determined in the previous step (Table 2.3) to obtain the priorities of the alternatives which constitutes the last section of the table. The last row shows the total impact of each alternative on environment, based on the chosen factors. The resulting values indicate that the Foamular

XPS is the most harmful roof material with the final score of 0.3177, while polyiso roof insulation is the most environment-friendly material with a score of 0.1426.

Table 2.5 Characteristics of roof materials and their environmental impacts

	Closed cell SPF	Foamular XPS	Mineral wool board	Polyiso roof insulation
Global warming (kg)	34.3	60.8	7.94	2.58
Acidification (Mol H$^+$)	1.073	1.78	2.75	0.0098
Ozone depletion (kg CFC-11)	167	36300	118	94.7
Smog creation (kg O_2)	0.267	0.208	0.254	0.18
Eutrophication (kg N)	1.33	9.85	9.4	14
Primary energy demand (MJ)	136.7	80.7	98.7	63.8
Waste to disposal (kg)	0.011	0.857	5.12	3.7
Water demand (kg)	1065	37.9	1.27	172
Normalized				
Global warming	0.3247	0.5756	0.0752	0.0244
Acidification	0.1912	0.3171	0.4900	0.0017
Ozone depletion	0.0046	0.9896	0.0032	0.0026
Smog creation	0.2937	0.2288	0.2794	0.1980
Eutrophication	0.0385	0.2848	0.2718	0.4049
Primary energy demand	0.3598	0.2124	0.2598	0.1679
Waste to disposal	0.0011	0.0885	0.5285	0.3819
Water demand	0.8345	0.0297	0.0010	0.1348
Priorities				
Global warming	0.0303	0.0537	0.0070	0.0023
Acidification	0.0357	0.0593	0.0916	0.0003
Ozone depletion	0.0005	0.1027	0.0003	0.0003
Smog creation	0.0200	0.0156	0.0191	0.0135
Eutrophication	0.0020	0.0145	0.0138	0.0206
Primary energy demand	0.0986	0.0582	0.0712	0.0460
Waste to disposal	0.0001	0.0106	0.0632	0.0457
Water demand	0.0862	0.0031	0.0001	0.0139
	0.2734	**0.3177**	**0.2664**	**0.1426**

Therefore, it is obvious that there is a considerably high potential for developing the sustainability by reducing the ecological effect of the insulation products and develop the buildings' environmental ratings. Taking this into consideration that generation is the primary procedure and that the utilization of raw materials and energy are the primary topics to work on, one could think of measures focusing on [40]:

- the utilization of sustainable energy resources for the power required at the production procedure, or on area-off-site, or biomass, such as green electric,
- the end-of-life administration of construction materials and the calculation of waste flows at the generation procedures, covering the recovery, recycling, and reuse potential,
- the industrial infrastructure's upgrading (improved monitoring, refurbishment, and energy consumption's control, Energy Management Systems' implementation (for example, ISO50001),
- decreasing the transportation emissions by fostering the utilize of biofuels for the vehicles and preferring locally extracted raw materials.
- recycling and reuse of construction materials which is necessary in order to diminish the complicated power in constructions. The recycled insulating material's utilization could lead to savings of more in construction's complicated energy.

2.5 Conclusions

As the public knowledge of the environment is growing and modern structures are becoming increasingly power effective, the focus is on the use of environmentally friendly construction products. Emotions and perceptions often drive human choices. There is therefore a powerful inclination to prefer "natural" building goods to synthetic ones, particularly in the case of heat insulation. Life-cycle evaluation has made it possible to broaden the significance of the term "eco-friendliness" by providing scientists and construction planners with a purpose decision-making tool to define the ecological effect of building materials, construction parts and constructions all together. Decreasing energy expenditure in buildings is essential to lowering or restricting the construction sector's adverse environmental effects.

Implementation of insulation materials is an efficient way to reduce energy consumption in structures related to heating and cooling. Building insulation adoption in the globe was mainly motivated by construction codes and norms, with little attention being paid to the ecological advantages of more sophisticated insulation products. Advances in the sector have produced accessible products for building insulation that are both energy-efficient and better for the ecology, with reduced effects on the environment throughout life.

For the purpose of this study, life-cycle information was gathered from the Environmental Product Declaration and Health Product Declaration of the roof insulation material manufacturers to evaluate the immediate environmental effects. To supplement this data, parameters for chosen insulating materials were evaluated by experts. In this chapter, the environmental databases and AHP evaluation methodologies were utilized to assess the effects in the insulation products' life cycle. For roof design, the most important environmental advantages of thermal insulation materials are discovered in the case of polyiso roof insulation.

As part of the conclusions of the analysis and related literature, we can specify the following:

- In the case of insulating materials, their utilization is essential to accomplish next energy conservations at the maximum grade the life-cycle evaluation methodology displays that the operating phase is responsible for approximately 80% of the overall ecological effect. Taking this into consideration, utilization of insulating materials is still necessary, even if their pollution determiners are not zero.

- Within the framework of maintainable constructions, the life-cycle evaluation methodology appears to be a beneficial technique for the assessment of ecological expenses of constructions operation and performance. The basic life-cycle evaluation can be improved as a technique of ecological maintainability evaluation, and partly can give rise to the social topics assessment, in terms of health risks. Life-cycle cost calculation, economical alteration to the fundamental life-cycle assessment, can be utilized to assess financial topics linked with construction.

- The holistic point of view of this study makes the life-cycle evaluation methodology proper for the assessment of overall ways of maintainable improvement. The notion of evaluating the constructions' energy performance is in demand to decrease the amount of energy depleted. This notion occurs of both financial (greater fuel costs) and environmental matters such as the ecological conservation from reverse utilization impacts. From an ecological perspective it is too significant to implement and develop novel, efficient products and technologies with the generation process characteristics optimized, decrease energy requisition and also the expenditure of feedstocks.
- Appropriate materials utilized in the building industry can heavily affect the comfort of buildings, and thus the health of the occupiers. For the building industry, the implementation of the ecological evaluation methodologies seems to be justified due to its extreme energy and material expenditure. As insulating materials are utilized to decrease the energy expenditure in constructions, these materials are usually noted to be "energy performance" materials. The utilization of insulating materials can affect human comfort.
- It is still significant to take into account the ecological effects induced by their generation. The generation of the insulating materials, such as every generation process, contains an expenditure of raw materials and energy as well as the emit of polluters. The significance of the ecological investigation of insulating materials is very important within the framework of maintainable constructions.

In terms of insulating materials' ecological effect, it can be concluded that the utilization of sources reaches in the heart of the materials' effect. It is therefore requisite to support innovator solutions and the utilization of finest existing methods in the generation operations, in order to diminish the consuming of the natural, finite sources. In that sense, encouraging the utilization of sources that are available locally is one of the most significant gauges to diminish transportation emissions and expanses. Moreover, energy expenditure in the generation operations is demonstrated to be viable for the primary part of carbon dioxide emissions.

References

1. https://www.efficientbuildings.com/gallery/realities-heat-loss (26.09.2019).

2. Sismana, N., Kahyab, E., Arasc, N., Arasa, H. (2007). Determination of optimum insulation thicknesses of the external walls. *Energy Policy*, 35(10), 5151–5155.

3. Balo, F. (2011). Energy and economic analyses of insulated exterior walls for four different city in Turkey, *Energy Educ. Sci. Technol. Part A: Energy Sci. Res.*, 26(2), 175–188.

4. Spanaki, A., Kolokotsa, D., Tsoutsos, T., Zacharopoulos, I. (2014). Assessing the passive cooling effect of the ventilated pond protected with a reflecting layer. *Appl. Energy*, 123, 273–280.

5. Tong, S., Li, H. (2014). An efficient model development and experimental study for the heat transfer in naturally ventilated inclined roofs. *Build Environ*, 81, 296–308.

6. Synnefa, A., Santamouris, M., Akbari, H. (2007). Estimating the effect of using cool coatings on energy loads and thermal comfort in residential buildings in various climatic conditions. *Energy Build*, 39, 1167–1174.

7. Balo, F. (2011). Castor oil-based building materials reinforced with fly ash, clay, expanded perlite and pumice powder. *Ceramics Silikaty*, 55(3), 280–293.

8. Al-Sanea, S. A. (2002). Thermal performance of building roof elements. *Build Environ.*, 37, 665–675.

9. D'Orazio, M., Di Perna, C., Di Giuseppe, E. (2010). The effects of roof covering on the thermal performance of highly insulated roofs in Mediterranean climates. *Energy Build*, 42, 1619–1627.

10. Balo, F. (2015). Characterization of green building materials manufactured from canola oil and natural zeolite. *J. Mater. Cycles Waste Manag.*, 17, 336–349.

11. https://greenteamuk.net/room-in-roof-insulation/(26.09.2019).

12. https://www.reportsnreports.com/contacts/requestsample.aspx?name=2459236 (26.09.2019).

13. www.holzfragen.de, Sachverständigenbüro für Holzschutz Hans-Joachim Rüpke/Dr. Ernst Kürsten Hannover.

14. Balo, F. (2015). Feasibility study of "green" insulation materials including tall oil: environmental, economical and thermal properties. *Energy Buildings*, 86, 161–175.

15. Ekincioglu, O., Gurgun, A. P., Engin, Y., Tarhan, M., and Kumbaracibasi, S. (2013). Approaches for sustainable cement production–A case study from Turkey. *Energy Buildings*, 66, 136–142.

16. Dylewski, R., and Adamczyk, J. (2012). Economic and ecological indicators for thermal insulating building investments, *Energy Buildings*, 54, 88–95.

17. Shrestha, S. S., Biswas, K., and Desjarlais, A. O. (2014). A protocol for lifetime energy and environmental impact assessment of building insulation materials. *Environ. Impact Assess. Rev.*, 46, 25–31.

18. Asif, M., Muneer, T., and Kelley, R. (2007). Life cycle assessment: A case study of a dwelling home in Scotland. *Building Environ.*, 42(3), 1391–1394.

19. Crawford, R. (2011). *Life Cycle Assessment in the Built Environment*. Taylor and Francis.

20. Ramesh, T., Prakash, R., and Shukla, K. K. (2010). Life cycle energy analysis of buildings: An overview. *Energy Buildings*, 42(10), 1592–1600.

21. Kotaji, S., Schuurmans, A., and Edwards, S. (eds.). (2003). *Life-Cycle Assessment in Building and Construction: A State-of-the-Art Report of Setac Europe Setac.*

22. Mackey, C. O., Wright, L. T. (1944). Periodic heat flow-homogeneous walls or roofs. *ASHVE Trans.*, 50, 293.

23. Rees, S. J., Spitler, J. D., Davies, M. G., Haves, P. (2000). Qualitative comparison of North American and UK. Cooling load calculation methods. *Int. J. Heating Ventilating. Air Cond. Refrigeration Res.*, 6(1), 75–99.

24. Balo, F. (2017). Evaluation of ecological insulation material manufacturing with analytical hierarchy process (Ekolojik yalýtým malzemesi üretiminin analitik hiyerarşi prosesi ile değerlendirilmesi), *J. Polytechnic (Politeknik Dergisi)*, 20(3), 733–742.

25. Ilgin, M. A., Gupta, S. M. (2010). Environmentally conscious manufacturing and product recovery (ECMPRO): A review of the state of the art. *J. Environ. Manag.*, 91(3), 563–591. doi:10.1016/j.jenvman.2009.09.037. PMID 19853369.

26. EPA NRMRL Staff (6 March 2012). Life cycle assessment (LCA). EPA.gov. Washington, DC. EPA National Risk Management Research Laboratory (NRMRL). Archived from the original on 6 March 2012. Retrieved 17 March 2020.

27. Jonker, G., Harmsen, J. (2012). Creating design solutions (§ goal definition and scoping). In: *Engineering for Sustainability*. Amsterdam, NL: Elsevier. Chapter 4, pp. 61–81, esp. 70. doi:10.1016/B978-0-444-53846-8.00004-4. ISBN 9780444538468. Retrieved 8 December 2019.

28. Life Cycle Assessment (LCA) Overview. https://sftool.gov. Retrieved 17 March 2020.

29. Rebitzer, G., et al. (2004). Life cycle assessment. Part 1: Framework, goal and scope definition, inventory analysis, and applications. *Environ. Int.*, 30(5), 701–720. doi:10.1016/j.envint.2003.11.005. PMID 15051246.

30. Finnveden, G., Hauschild, M. Z., Ekvall, T., Guinée, J., Heijungs, R., Hellweq, S., Koehler, A., Pennington, D., Suh, S. (2009). Recent developments in Life Cycle Assessment. *J. Environ. Manage.* 91(1), 1–21. doi:10.1016/j.jenvman.2009.06.018. PMID 19716647.

31. ISO 14044:2006. ISO. Retrieved 17 March 2020.

32. Flysjö, A, Cederberg, C., Henriksson, M., Ledgard, S. (2011). How does co-product handling affect the carbon footprint of milk? Case study of milk production in New Zealand and Sweden. *Int. J. Life Cycle Assess.*, 16(5), 420–430. doi:10.1007/s11367-011-0283-9.

33. Rich, B. D. (2015). Future-proof building materials: A life cycle analysis. In: Gines, J., Carraher, E., Galarze, J. (eds.). *Intersections and Adjacencies. Proceedings of the 2015 Building Educators' Society Conference.* Salt Lake City, UT: University of Utah. pp. 123–130.

34. Saling, P., ISO Technical Committee 207/SC 5 (2006). ISO 14044: Environmental management—Life cycle assessment, Requirements and guidelines (Report). Geneve, CH: International Organisation for Standardisation (ISO). Retrieved 11 December 2019.

35. Saling, P., ISO Technical Committee 207/SC 5 (2006). ISO 14040: Environmental management—Life cycle assessment, Principles and framework (Report). Geneve, CH: International Organisation for Standardisation (ISO).

36. Curran, M. A. Life Cycle Analysis: Principles and Practice (PDF). Scientific Applications International Corporation. Archived from the original (PDF) on 18 October 2011. Retrieved 18 March 2020. 13. Scientific Applications International Corporation (May 2006). Life cycle assessment: principles and practice (PDF). p. 88. Archived from the original (PDF) on 23 November 2009.

37. Singh, S., Bakshi, B. R. (2009). Eco-LCA: A tool for quantifying the role of ecological resources in LCA. *International Symposium on Sustainable Systems and Technology*: 1–6. doi:10.1109/ISSST.2009.5156770. ISBN 9781424443246.

38. Griffin, C. (2015). Halogenated Flame Retardants and Insulation Selection: The Role of Transparency. JMRoofing.news.

39. Goepel, K. D. (2013). Implementing the analytic hierarchy process as a standard method for multi-criteria decision making in corporate enterprises: A new AHP Excel template with multiple inputs. *Proceedings of the International Symposium on the Analytic Hierarchy Process*, pp. 1–10.

40. Giama, E., Papadopoulos, A. M. (2015). Assessment tools for the environmental evaluation of concrete, plaster and brick elements production, *J. Cleaner Prod.*, 1–11, doi: 10.1016/j.jclepro.2015.03.006.

Chapter 3

Mathematical Radiation Models for Sustainable Innovation in Smart and Clean Cities

Figen Balo[a] and Lutfu S. Sua[b]

[a]*Department of Industrial Engineering,*
Fırat University, Turkey
[b]*Department of Management and Marketing,*
Southern University, Louisiana, USA

fbalo@firat.edu.tr, lutsua@gmail.com

Sustainable energy has started to attract many researchers due to the ecological imbalance as a result of fossil-based energy sources. The use of renewable energy sources instead of fossil-based resources is a critical choice in preserving the ecological balance when making smart city designs. When all sorts of renewable energy sources are explored, solar energy appears to be the most applicable resource in smart and clean city designs. Solar energy is a source that can be obtained in most of the cities more or less. It is also a clean and relatively low-cost alternative without carbon emission in the production process and it can be stored. Even the intensity of the solar energy in a city is low and it can be transformed into mechanical and electrical (e.g., photovoltaic) energies with appropriate efficiency. Thus, solar energy is the optimum choice for smart and clean city designs.

Climate Change and Pragmatic Engineering Mitigation
Edited by Jacqueline A. Stagner and David S.-K. Ting
Copyright © 2022 Jenny Stanford Publishing Pte. Ltd.
ISBN 978-981-4877-97-8 (Hardcover), 978-1-003-25658-8 (eBook)
www.jennystanford.com

Solar energy's potential in meeting energy requirements stems from the need for efficient use of scarce resources worldwide and environmental considerations. Recently, all countries are in search of ways of increasing renewable energy's share among other energy generation methodologies to provide energy independence. However, investment in all sorts of renewable energy requires in-depth feasibility studies that rely heavily on available data. The lack of solar radiation data imposes a barrier in this sense. Thus, there are many studies involving data estimation model development. One of the challenges is the fact that geographic and climatic characteristics of the region that makes a significant difference in system efficiency. In this paper, to design solar systems for optimum performance within specific climatic conditions, sun energy potency for two major cities in the Black Sea region is provided through a comparative analysis. Taking these conditions into account and using the appropriate radiation models, all radiation values and consequently the total radiation values are calculated for both regions under investigation. Simulation analysis of the two provinces revealed that the best max total irradiation value was obtained as 6.6135 for Samsun.

3.1 Introduction

Sun energy is increasing its popularity due to various reasons such as environmentally hazardous waste caused by conventional energy resources and the growing energy demand of the increasing global population. The limited amount of fossil-based sources is a significant issue that leads the research efforts towards increasing the efficiency of renewable energy sources so they can make up for the energy shortage. At this point, it is vital to investigate both the technical and financial feasibilities of solar systems before investing a considerable amount of time, capital, and human hours. The sun irradiation is among the main indicators when determining the potential of solar energy in a specific region. The amount of net absorbed global sun irradiation map and Earth's radiation budget are shown in Figs. 3.1 and 3.2, respectively.

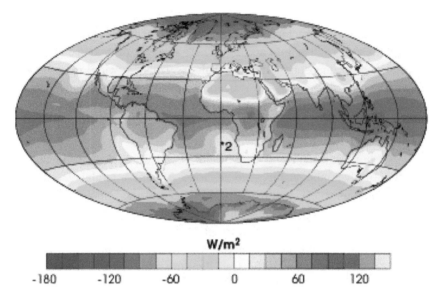

Figure 3.1 The amount of net absorbed global solar radiation [1].

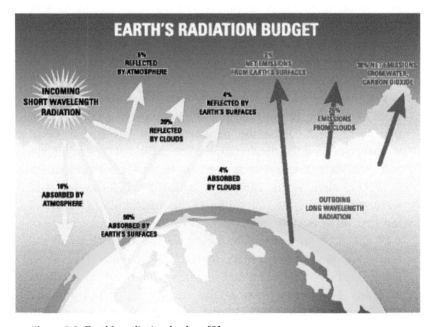

Figure 3.2 Earth's radiation budget [2].

As sun irradiation is heavily interdependent on the climatic and geographical characteristics of a location, feasibility studies need to take these issues into account. There is a lack of meteorological stations to measure radiation values. Thus, a wide range of mathematical models have been proposed by researchers to investigate these values. Many articles have also indicated that the artificial neural network methodology is superior to the empirical models [3–5].

Khorasanizadeh and co-workers [6, 7] studied six different quantitative models for several cities. The first model was an exponential, the second one a polynomial and the remaining ones were cosine and sine functions. The mean square error of the six models was within the range of 1.26 and 0.72 MJ/m^{2}day, and the mean absolute percent error was reported to vary between 5.72% and 3.38%. For Dezful, Iran, Behrang, and co-workers studied radial base function system and multilayer perceptron network. The six integrations of the characteristics used wind velocity, relative moisture, day number, sunshine period, evaporation, and average air temperature as inputs. For the analysis, 214 days of data were utilized. The value of MAPE was reported to vary between 5.21% and 22.88%. [8]. Kasra and co-workers used sunshine fraction models using nine years. Four years of information were utilized for testing. The RMSE value changed within the range of 1.18 and 1.1 MJ/m^{2}day [9]. Li and co-workers were looking for 12 non-sunshine fraction models (NSDF). In the first three models, the fog parameters, the mean dew point temperature, and the precipitation were combined. The minimum and maximum temperatures were used in the other models. Data were obtained for 2552 days for testing. 2921 days of data were obtained for the calibration of the models. RMSE values of the first three models were reported to vary within the range of 6.24 and 5.18 MJ/m^{2}day. The range for the error terms of the remaining models were 3.05 and 2.52 MJ/m^{2}day [10]. Zang and co-workers [11] examined the same model for 35 locations in China by reducing two of the coefficients. MAPE and the RMSE terms for the selected locations changed in the ranges of 16.22%–4.33% and 1.88–1.10 MJ/m^{2}day.

For nine locations, Zhao and co-workers explored a linear model. The root mean square error was reported to vary from 1.72 to 5.24 MJ/m²day [12]. For 79 locations in China, Li and co-workers [13] performed an integrated modeling composed of cosine and sine functions using data of a 10-year period. RMSE varied in the range of 1.03 and 1.83 MJ/m²day while the mean square error value changed between 4% and 15.4%. Jamshid Piri and co-workers examined a modified modeling of the sunshine duration fraction for two locations and three models of the sunshine duration fraction (SDF) were used. They utilized the support vector regression method. The RMSE values were reported to be within the change of 3.70 and 2.14 MJ/m²day. For the kernel function, the lowest and highest temperature, sunshine period, and relative moisture parameters were chosen as entries [14]. Park and co-workers used a linear empirical model for 22 locations in South Korea [15]. For 25 sites, Manzano and co-workers assessed linear Angstrom-Prescott modeling in Spain. Data from more than 10 years is used for the calibration. With the exception of four locations, the RMSE changed in the range of 0.8 and 0.36 MJ/m²day [16]. Tang and co-workers were looking for a model introduced by Koike and Yang to predict daily solar radiation [17]. This hybrid modeling computed the global irradiation for clear sky and clear sky index. These models were composed of parameters such as sunshine duration, the thickness of the air pressure, the ozone layer, relative moisture, air temperature, surface height, and angstrom turbidity. For 97 weather stations in China, the radiation data received for the period of 1993–2000 were utilized to approve the hybrid modeling. The quadratic average error varied within the range of 1.3 and 0.7 MJ/m²day [18].

In Turkey, Ahmet Teke and co-workers examined cubic, linear, and quadratic empiric modeling for four provinces [19]. Li and co-workers assessed eight sunshine fraction modeling for four locations. Data for a period of 11 years were used for calibration purposes while 4 years of data were utilized to validate the model. RMSE was utilized as a statistical index and its value for the linear model was reported to vary from 1.26 to 0.72 MJ/m²day. The eight modeling's root mean square error ranged between 1.33 and 0.7 MJ/m²day [20]. Liu and co-workers examined three models without sunshine duration, three modified

SDF models, and 2SDF modeling. 1085 days of data were tested for calibration. Data for 701 days were utilized to validate the models. The quadratic mean error showed variation from 3.13 to 1.68 MJ/m^{2}day. For different seasons, they denoted that it was not necessary to reproduce coefficients [21]. Katiyar and co-workers researched the cubic, linear, and quadratic modeling for the estimate for the monthly average irradiation for four cities through the yearly data. The values were between 0.43 and 0.8 MJ/m^{2}day [22]. For 41 locations in China, Kevin et al. performed the linear Prescott-Angstrom modeling to forecast daily global solar radiation. Depending on diverse factors, these locations were divided into seven sun climatic regions and nine thermal climatic regions. They performed the ANN modeling through the use of inputs such as latitude, height, longitude, number of days, daily average temperature, and sunshine period fraction [23].

By using general features, Kevin and co-workers achieved the highest performance. Data is generated by calculating the average of data obtained from diverse sites, reducing the changes' amplitude in sun irradiation and other features [23, 24]. Gorka and co-workers provided a comparison of an ANN modeling, adaptive neuro-fuzzy inference system, and gene expression programming. A large number of observations (2855) from four stations were obtained for testing. 4420 observations were obtained for the model training. The modeling utilized the integrations of five characteristics (minimum and maximum air temperature, extraterrestrial irradiation, irradiation with a clear sky, and the number of days) as inputs. The quadratic mean error of the optimized GEP ranged from 3.31 to 3.49 MJ/m^{2}day. The mean square error of the corresponding optimized adaptive neuro-fuzzy inference system changed from 3.33 to 3.14 MJ/m^{2}day. The quadratic mean error of the optimized artificial neural networks when utilizing other three integrations as entries changed between 2.97 and 2.93 MJ/m^{2}day [25]. For four locations in Tunisia, Chelbi and co-workers examined five empirical models [26]. Chen and co-workers examined five sunshine fraction models. Data were obtained from each site for 35 years, 70% of which were studied to provide empiric coefficients. To analyze the dataset, 30% of the available information was utilized. The empiric coefficients were defined for each station. For Chaoyang,

RMSE was reported to vary from 1.98 to 2.73 MJ/m^{2}day [27]. For 69 sites, Zhou and co-workers tested six SDF modeling, and three SDF modeling were used to estimate on the monthly basis mean sun exposure in China. The height and width were added as characteristics in the replaced modeling. The coefficients were reproduced. The range of sunshine duration proportion was reported to be between 1.636 and 1.634 MJ/m^{2}day [28].

To estimate solar radiation, Sun and co-workers evaluated the effect of the auto-regressive moving mean modeling. In China, they examined the information for 20 years at two locations [29]. For seven sites, Hacer and co-workers examined five models for the period of the sunshine period to predict monthly mean irradiation in Turkey [30]. In 1 year, Ayodele and co-workers implied a function to represent the clarity index's dispersion. Utilizing 7 years, the coefficients defined daily sun exposure information. With the exception of October, the effectiveness was achieved every month. The RMSE ranged between 0.221 and 0.213 MJ/m^{2}day [31]. For six provinces, Khorasanizadeh and co-workers rated three average SDF modeling and three NSDF modeling for predicting the mean monthly global sun irradiation in Iran. In modeling with medium sunshine duration, the temperature and relative moisture were added as characteristics. Compared to models with sunshine period fraction, the quadratic average error values of all the models ranged between 0.82 and 0.47 MJ/m^{2}day [32]. Yao and co-workers investigated the 11 mathematical formulations of the hourly sun irradiation rate to daily solar radiation. They utilized information for 3 years to approve the modeling. Root mean square error showed variation within the range of 0.321 and 0.481 MJ/m^{2}h, corresponding to 88.33 and 142.22 W/m^2 [33]. Yao and co-workers studied 89 mean irradiation modeling. Through the use of different coefficient values, many models were used with the same mathematical formulations. New adjustment coefficients were derived for five models of the sunshine duration fraction in Shanghai [34]. For 11 Tibetan meteorological locations, Pan and co-workers examined the exponential modeling on temperature. The temperature variation was utilized as an entry. Information for 35 years was utilized to calibrate the modeling. Information for 5 years was utilized for analysis. The RMSE of the modeling was changed between 2.54 and 3.24 MJ/m²day for all of the stations [35]. For the Iranian province of Bandar Abass, Mohammadi and co-workers [36] used a wavelet convert algorithm and an assist vector machine. For the aim of developing the models, information for a period of 10 years was used. Parameters such as water vapor pressure, minimal and maximal ambient temperature differences, relative moisture, the sunshine period's proportion, extraterrestrial global sun irradiation, and the mean ambient temperature were utilized. RMSE values were reported to vary within the range of 1.81 and 1.79 MJ/m²day.

Antonio and co-workers developed a linear formula for correlating solar radiation with daily temperature fluctuations and the sunshine period's product utilizing the energy balance between the soil layer and neighboring atmospheric layer [37]. El-Sebaii and co-workers used three medium SDF modeling and NSDF modeling for predicting the mean monthly global sun irradiation. The parameters classified in models with medium sunshine fraction were temperature, relative moisture, and cloud cover. The data from 9 years were used to reproduce novel empiric coefficients. The RMSE values of the nine modelings were between 0.15 and 0.02 MJ/m²day [38, 39]. To select the relevant entry properties, Jiang defined Pearson correlation coefficients according to precedence relationships. As parameters, mean opaque sky cover, wind velocity, average temperature, precipitation, minimal-maximal temperature, daylight temperature, relative moisture, heating-cooling degree-days were selected [40]. Bakirci examined 60 empirical models designed to predict the global monthly and average daily sun exposure, with many of the forecasts having identical formulations, but with different regressive stable characteristics. All the same, according to the conclusions of numerous studies, these stable characteristics usually depend on the study fields [41]. To estimate the average sun exposure per hour, Janjai and co-workers received a satellite-based model. The RMSE between 3 p.m. and 9 a.m. varied between 10.7% and 7.5% [42]. For daily radiation, Wan Nik and co-workers analyzed six mathematical formulas of the hourly sun irradiation rate. The forecast was made for the monthly mean hourly irradiation. At three locations in Malaysia, information was used for 3 years to analyze the modeling. They concluded that the RMSE ranged between 26.49% and 8.22% [43].

For Akure in Nigeria, Adaramola was looking for six models with no sunshine period to forecast the long-term on the monthly basis average sun irradiation and the Angstrom-Page modeling. Relative moisture, precipitation, and environment temperature were utilized in models without the sunshine period. RMSE for the linear modeling changed between 8.25 and 4.78 MJ/m^{2}day [44]. In Turkey, Ozgoren and co-workers utilized the modeling of multi non-linear regression artificial neural networks to provide the most ideal independent properties for the entry sheet. They chose 10 parameters (day of the month, ground temperature, sunshine period, altitude, cloud cover, average ambient temperature, wind velocity, latitude, and minimal-maximal atmospheric temperature). The Marquardt–Levenberg optimization algorithm was utilized to develop artificial neural networks [45]. Senkal suggested an ANN modeling that uses height, latitude, longitude, terrain superficies temperature, and two different superficies emissivity as entries. The recent three features were defined utilizing satellite information. Data from 10 locations were used for 1 year to train the ANNs. The quadratic average error in the test and training phase was given as 0.16 and 0.32 MJ/m^{2}day [46]. For two locations, Jamshid and co-workers investigated two support vector regression models in Iran. The sunshine period, the minimal and maximal temperatures, and the relative moisture were used as inputs. RMSEs were determined to be between 4.47 and 1.63 MJ/m^{2}day [14]. For Iseyin, Lanre and co-workers implemented the adaptive neuro-fuzzy inference system and ANN in Nigeria. The inputs were sunshine period, minimal and maximal temperature values. Data from 6 years was used for model generation, while information from 15 years was used to analyze the models. In the test and developing phase, the RMSE ranged from 1.76 to 1.09 MJ/m^{2}day [47].

The majority of the energy consumption is being used for lightning, heating/cooling, and operation of all electrical devices. Thus, it is essential to investigate the sun energy potential and design of solar farms in cities where they will be applied. Solar energy is discontinuous and shows daily and seasonal variations.

Moreover, the amount of solar energy also varies depending on the atmospheric conditions where the city is located. Thus, while evaluating the availability of solar energy for the energy need of a city, it is important to use the models that can provide optimum solutions. The smart and clean cities designed in this manner can benefit from solar energy in the best possible way. This study intends to survey a selection of these models and apply the most appropriate ones for two selected regions in a comparative analysis.

3.2 Potential of Solar Energy and Climate of Trabzon and Samsun

Scarcity of equipment and cost levels of their maintenance are the barriers that limit the number of stations gauging sun irradiation. As a result, computing sun irradiation through the meteorological variables' use is a popular method in the literature [48, 49]. The sunshine period and land are very significant factors for feasibility studies of solar facilities. Thus, extensive analysis is vital to study and analyze the potential of solar energy, current facilities, as well as the climate of the region.

Seven geographic regions of Turkey have different climatic characteristics due to the geographical attributes and the altitude effect. Turkey horizontal irradiation is presented in Fig. 3.3.

This study focuses on the Black Sea Climatic regions which one of the five climatic regions in Turkey. The Black Sea region has an a climate with high and evenly distributed rainfall all over the year. The summers are warm and humid and the winters are cool and humid on the coast. The coast receives the most rainfall.

The two largest cities in the region both in terms of population and development are chosen to investigate the solar potential of the region. Both Trabzon and Samsun cities that are located on the coast of the Black Sea, present the same climatic characteristics. The mean sun irradiation function, phase shift, latitude, and radiation function frequency levels for the two cities under investigation are presented in Table 3.1.

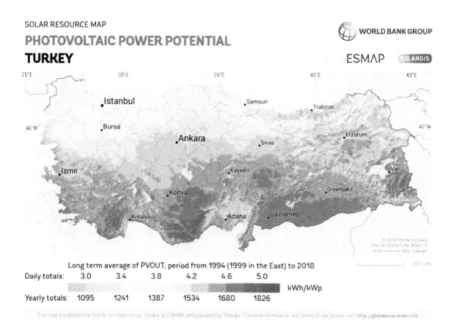

SOLAR RESOURCE MAP

PHOTOVOLTAIC POWER POTENTIAL

TURKEY

WORLD BANK GROUP

ESMAP

Long term average of PVOUT, period from 1994 (1999 in the East) to 2018

Daily totals: 3.0 3.4 3.8 4.2 4.6 5.0

kWh/kWp

Yearly totals: 1095 1241 1387 1534 1680 1826

Figure 3.3 Turkey horizontal irradiation.

Table 3.1 Irradiation levels

	FKI	Latitude	FGI (MJ/m^{2}day)	I_{avg} (MJ/m^{2}day)
Trabzon	13.2	41.00	5.47	9.85
Samsun	5.94	41.17	6.73	10.3

*FGI, irradiation function periodicity; FKI, irradiation function phase shift; I_{avg}, average daily total irradiation.

Sun irradiation maps for Trabzon and Samsun are shown in Fig. 3.4. The following section provides a comparative investigation conducted through a Matlab simulation program as an effort to calculate the solar potential and the characteristics of these cities.

This study aims to assess the solar energy potential of a climatic region through the use of quantitative models and simulation software. Climatic differences among different regions make it difficult to use a standard model for every region. Thus, region-specific models need to be developed. The first purpose of this study is obtaining such a model for the region under investigation.

The second problem in determining the solar potential is the lack of data. The second purpose of the study at hand is generating such data through the use of special simulation software.

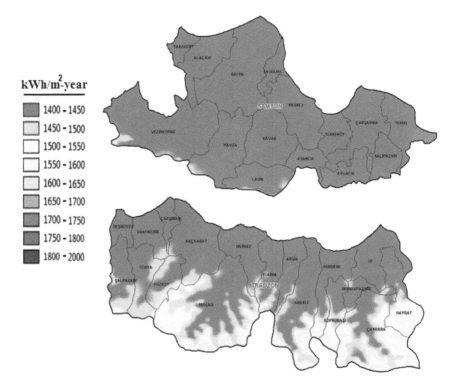

Figure 3.4 Sun irradiation maps for Trabzon and Samsun.

The chapter is organized under three sections. The Introduction section presents the existing literature and climatic characteristics of the region. The second section outlines the theory and model development for both the horizontal surfaces and inclined surfaces, followed by the calculations obtained from the simulation software. Finally, the last section presents the solar potential of the region through the numerical data obtained from the analysis.

3.3 Theory and Calculations

3.3.1 Horizontal Superficies

3.3.1.1 Daily total sun irradiation

There are numerous models formulated by the researchers for the purpose of finding out the amount of solar irradiation in different geographical locations of the world. The goal of sun irradiation models is to obtain solutions while considering the specific regions' characteristics. Dogniaux and Lemoine developed a mathematical sun irradiation model for each of the months [50]. Bahel introduced a mathematical model by modeling data obtained from 48 stations worldwide [51]. Jain developed eight different solar radiation models to use for diffuse and global solar radiation estimation for two different provinces in two countries [52]. Sfetsos and Coonik used ANN for future radiation value estimation [53].

On a given day, total sun irradiation on horizontal superficies can be computed by the following formulation [54]:

$$I = I_{avg} - \text{FGI}\cos\left[\frac{2\pi}{365}(n + \text{FKI})\right], \tag{3.1}$$

where FGI is irradiation function periodicity, FKI irradiation function phase shift, n days, and I_{avg}, yearly mean of daily total irradiation.

3.3.1.2 Daily diffuse sun irradiation

Daily total diffuse sun irradiation on a horizontal superficies can be defined utilizing Eq. 3.2 [55]:

$$Iy = I(1 - B)^2(1 + 3B^2) \tag{3.2}$$

where B is the transparency index and I_o out of atmosphere irradiation.

3.3.1.3 Total momentary sun irradiation

Momentary total sun irradiation on horizontal superficies can be calculated utilizing Eq. 3.3 [56, 57]:

$$I_o = \frac{24}{\pi} I_s \big(\cos(e)\cos(d)\sin(w_s) + w_s \sin(e)\sin(d) \big) f, \qquad (3.3)$$

where e is the latitude angle, d declination, I_s (W/m²), solar constant, f solar constant correction factor, and w_s the sunrise hour angle, computed using equations and tables.

Irradiation is computed using the following formula [55]:

$$I_{ts} = A_{ts} \cos\left[\frac{\pi}{t_{gi}} (t - 12) \right], \qquad (3.4)$$

where t_{gi} is the imaginary day length and A_{ts} sun irradiation.

3.3.1.4 Diffuse and direct sun irradiation

Momentary direct and diffuse sun irradiation amount on horizontal superficies can be defined through Eqs. 3.5 and 3.6 [55–57]:

$$I_{ys} = A_{ys} \cos\left[\frac{\pi}{t_g} (t - 12) \right], \qquad (3.5)$$

where A_{ys} is function frequency.

$$I_{ds} = I_{ts} = I_{ys} \qquad (3.6)$$

3.3.2 On Inclined Superficies, Computing Sun Irradiation Intensity

3.3.2.1 Direct momentary sun radiation

Direct momentary sun radiation on surfaces with various angles (30, 60, and 90) is computed utilizing the below formulation as it follows [62]:

$$I_{bc} = I_b R_b \qquad (3.7)$$

$$R_b = \frac{\cos\theta}{\cos\theta_z} \tag{3.8}$$

$$\cos\theta_z = \sin d \sin e + \cos d \cos e \cos w \tag{3.9}$$

$$\cos\theta = \sin d \sin(e - \beta) + \cos d \cos(e - \beta)\cos w \tag{3.10}$$

3.3.2.2   Diffuse sun radiation

Diffuse radiation can be defined using the formulation as follows [62]:

$$I_{ye} = R_y I_{ys} \tag{3.11}$$

Conversion factor R_y for diffuse irradiation can be determined by the following formulation [62]:

$$R_y = \frac{1 + \cos(a)}{2}, \tag{3.12}$$

R_y parameter ensures the surface's slope. R_y value is 0.5 for vertical surface ($a = 90°$). For vertical superficies, R_y is 0.5. Using this value, diffuse irradiation values for 30°, 60°, and 90° surfaces can be computed.

3.3.2.3   Momentary reflecting irradiation

Reflecting irradiation [55–57] is computed through the equation as follows:

$$I_{ya} = I_{ts} p \frac{1 + \cos(a)}{2} \tag{3.13}$$

The reflecting rate to the environment is displayed with ρ and utilized with a mean value of $\rho = 0.2$.

3.3.2.4   Total sun irradiation

Total irradiation on inclined superficies is computed [57] using the equation as follows:

$$I_t = I_{de} + I_{ye} + I_{ya} \tag{3.14}$$

3.3.3   Simulation Results

Figure 3.5 shows the values of (a) variation of current yearly average sun irradiation levels over a 24 h cycle, (b) variation in current yearly diffuse sun irradiation values per hour, (c) variation of the current annual direct solar radiation levels for a 24 h cycle.

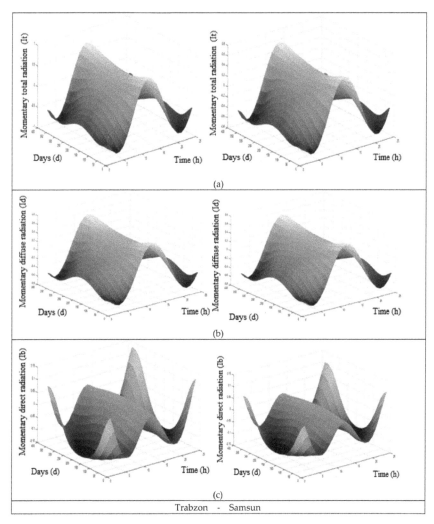

Figure 3.5 On horizontal superficies, annual sun irradiation variation over 24 h.

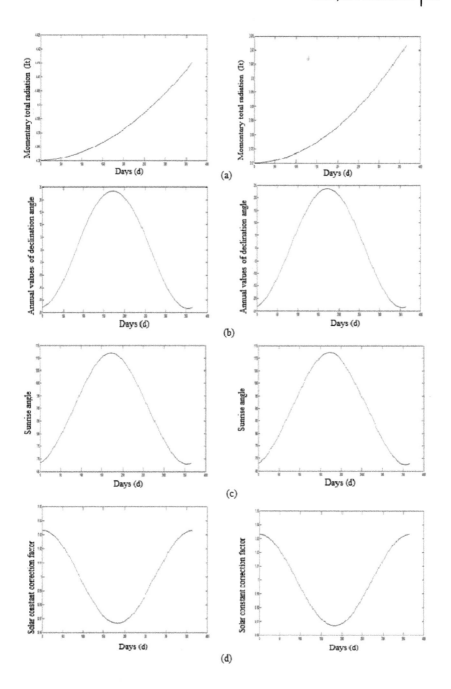

(a)

(b)

(c)

(d)

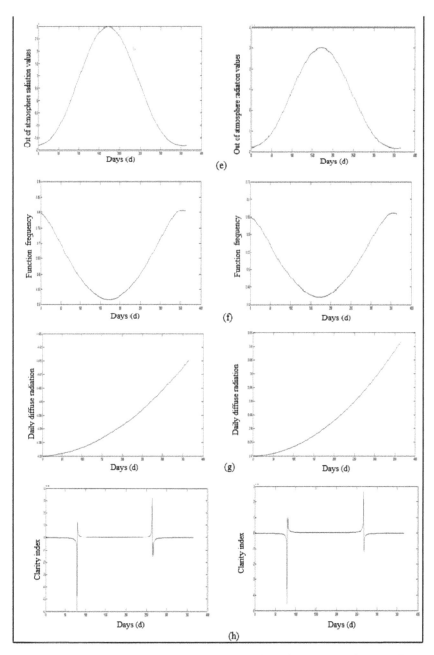

Figure 3.6 The sun irradiation on horizontal surfaces for Trabzon and Samsun.

Figure 3.6 shows daily variations of

(a) total daily sun irradiation;
(b) sunrise declination angle ;
(c) sunrise angle (hourly);
(d) correction factor;
(e) atmospheric irradiation;
(f) function periodicity graph (A_{ys});
(g) diffuse sun irradiation (A_{ts});
(h) horizontal surfaces transparency index (B).

Daily momentary direct radiation levels for the three angles can be observed in Fig. 3.7. The maximum values for the three angles are determined on day 355 at 12:00 and the minimum value at 03:00 on the same day.

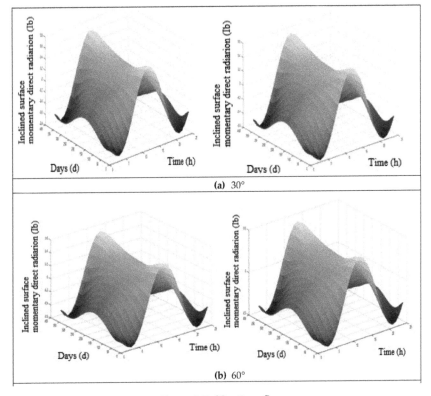

(a) 30°

(b) 60°

Figure 3.7 (*Continued*).

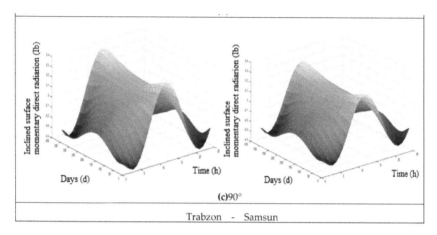

(c)90°

Trabzon - Samsun

Figure 3.7 Yearly momentary direct irradiation levels (24 h).

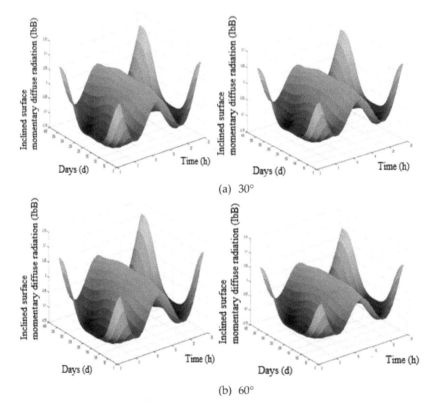

(a) 30°

(b) 60°

Figure 3.8 (*Continued*).

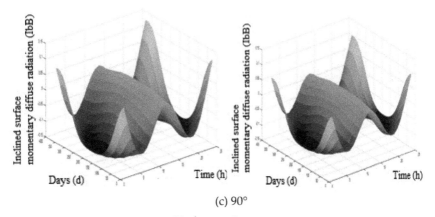

(c) 90°

Trabzon - Samsun

Figure 3.8 Yearly momentary diffuse irradiation levels.

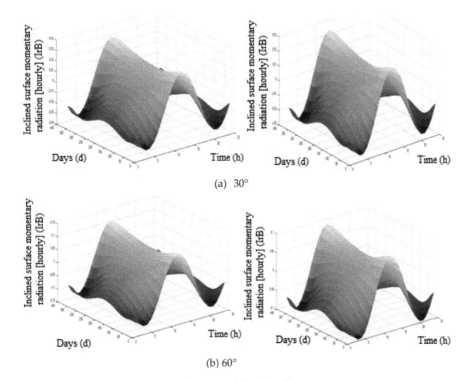

(a) 30°

(b) 60°

Figure 3.9 (*Continued*).

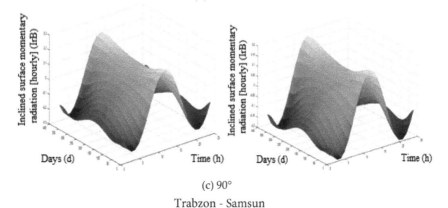

(c) 90°

Trabzon - Samsun

Figure 3.9 Yearly total momentary irradiation.

Figure 3.8 provides the yearly momentary diffuse irradiation values for the inclined superficies. The total momentary sun irradiation levels annually are displayed in Fig. 3.9.

3.4 Computational Conclusions

Resulting from the previous sections, the true potency of the provinces under investigation can be computed using calculations of the solar characteristics given in Table 3.2.

Table 3.2 Sun irradiation characteristics

Characteristics		Trabzon	Samsun
Total irradiation (kW/m²)	Max	4.4187	6.6135
	Min	4.3800	3.5700
Declination angle	Max	23.4898	23.4498
	Min	−23.4198	−23.4498
Sunrise hour angle	Max	113.1243	110.0186
	Min	68.9678	60.9814
Out-of-Atmosphere irradiation (kW/m²)	Max	300000	278010
	Min	−170000	−176900
Transparency Index	Max	0.3567	0.1630
	Min	−0.0060	−0.0044

Characteristics		Trabzon	Samsun
Total diffuse irradiation (kW/m^2)	Max	4.4146	8.8522
	Min	4.3800	8.4200
Function frequency (MJ/m^2day)	Max	0.7967	1.5075
	Min	0.5246	1.0104
Momentary Total Radiation (kW/m^2)	Max	0.7568	1.5075
	Min	−0.9754	−1.5044
Momentary Diffuse Radiation (kW/m^2)	Max	0.7654	1.4253
	Min	0.4710	0.9028
	Max	0.7567	1.4253
	Min	−0.8654	−1.4165
Momentary direct radiation (kW/m^2)	Max (30°)	0.6123	1.3299
	Min (30°)	−0.7654	−1.3216
	Max (60°)	0.5687	1.0690
	Min (60°)	−05412	−1.0624
	Max (90°)	0.3350	0.7127
	Min (90°)	−0.3983	−0.7082
Momentary diffuse radiation (kW/m^2)	Max (30°)	0.0125	0.2457
	Min (30°)	−0.1367	−0.2545
	Max (60°)	0.0210	0.2461
	Min (60°)	−0.1189	−0.2549
	Max (90°)	0.0436	0.2458
	Min (90°)	−0.0127	−0.2545
Momentary reflecting radiation (kW/m^2)	Max (30°)	0.0297	0.0606
	Min (30°)	−0.0385	−0.0605
	Max (60°)	0.1325	0.2261
	Min (60°)	−0.1488	−0.2257
	Max (90°)	0.2897	0.4522
	Min (90°)	−0.2921	−0.4513

The sun irradiation values on various surfaces can be obtained through MATLAB software. In terms of the system efficiency, the total solar radiation value in the region should be at least

4.5 kWh/m^2 per day and the annual sunshine duration should be more than 6 h/day. Calculations indicate that solar system potential in both provinces corresponds to the required values.

In the production of electrical energy with photovoltaic systems, semiconductor materials called photovoltaic cells convert sunlight directly into electrical energy. In these systems, solar energy can be transformed into electrical energy with efficiency between 5 to 30 percent depending on the structure of the solar cell. When designing a solar power plant, the decision is made by evaluating the basic criteria such as solar energy potential of the region, local climate condition, and land structure. One of the main conditions affecting the efficiency of solar power plants is the solar energy potential of the region. The solar radiation level of the region is directly proportional to the amount of energy produced. Solar energy can be converted into electrical energy with efficiency between 5% and 20% depending on the structure of the solar panel.

Usually solar radiation calculations are made on the day corresponding to the average monthly declination angle, rather every individual day of a month. Solar radiation calculations are also made according to solar time. The clock system, in which the sun azimuth angle is 0°, is called sundial (local time). Sunrise hour angle is the angle between the longitude of the sun's rays (called the sun's longitude) and the longitude of the considered location. Taking these angles into account in the provinces analyzed in Table 3.2, all radiation values and consequently the total radiation values are calculated as max and min. In this study, when the two provinces are compared, the best max total irradiation value was obtained as 6.6135 for Samsun.

Table 3.2 shows that the amount of total radiation in Samsun reaches considerably higher levels (6.6135) than its counterpart (4.4187) while the range of maximum and minimum values is wider for Samsun which reflects a high level of variation. An integral part of designing the solar systems is comparing the forecasted values with the real ones. System performances depend on numerous parameters. Optimum system design depends heavily on the use of realistic radiation values. This chapter provides a reference for selecting the most performance photovoltaic panel by depending on the actual sun irradiation values determined for the best performing PV system plan. The sun irradiation values are assessed to be at reasonable effectiveness standards to plan a PV system for a smart and clean city.

Increasing competitive power, growing economies, increasing the quality of life in societies are directly related to the technological development levels. Technological developments can be achieved with sufficient, continuous, and clean energy. However, the initial installation costs per kW of solar power plants are 3 times the costs of natural gas and petroleum power plants, 2 times the coal power plant, 1.5 times the hydroelectric power plant, and 1.2 times the wind power plant. For this reason, it is important to establish solar farms in suitable areas. For solar farms designed with the use of many expensive materials, it is very important that the place with appropriate total irradiation value is analyzed in detail by considering many parameters and then all positive and negative aspects of the design are evaluated.

Installation conditions depending on the radiation values of the place where solar farms are installed are directly related to the amount of electrical energy that can be obtained. As the total amount of solar radiation that can be obtained in connection with the geographical location of the current region decreases, the amount of energy that can be obtained from the panel in the system to be established with the same type of panels becomes less. This can be thought of as a lower amount of energy that can be obtained from a solar panel operating in a place with a lower temperature. For this reason, parameters related to solar radiation obtained from the city of Samsun offer a little more convenience for solar power plant installation in this city compared to Trabzon. When all data and calculated parameters are included in the study, the total irradiation value obtained is one of the most accurate indicators for the efficiency that can be obtained from the design when evaluated together with other conditions. Although the maximum total irradiation value for Trabzon province (4.4187 kWh/m^2) is close to the usable limit, it is lower than the maximum total irradiation (6.6135 kWh/m^2) value obtained from Samsun. Turkey is located in a very suitable geographical location in terms of solar radiation. This potential is higher in the solar industry than in the world's leading countries. According to the data of the General Directorate of Renewable Energy, the total solar radiation intensity is 1,311 kWh/m^2 annually and the total annual sunshine duration is 2,640 h. Solar Energy Potential Atlas Turkey (GEPA) shows that daily radiation values are at least 3.29 kWh / m^2, and the maximum value is 5.48 kWh/m^2. Results show that the total irradiation conditions for Samsun are quite good in comparison with the national average.

References

1. Image courtesy Dennis Hartmann, University of Washington via NASA. http://www.fondriest.com/environmentalmeasurements/ parameters/weather/photosynthetically-active-radiation/.

2. keywordteam.net.

3. Qazi A, Fayaz H, Wadi A, Raj RG, Rahim NA, Khan WA. The artificial neural network for solar radiation prediction and designing solar systems: a systematic literature review. *J Clean Prod* 2015;104:1–12.

4. Teke A, Yýldýrým HB, Çelik Ö. Evaluation and performance comparison of different models for the estimation of solar radiation. *Renew Sustain Energy Rev* 2015;50:1097–1107.

5. Piri J, Kisi O. Modelling solar radiation reached to the Earth using ANFIS, NNARX, and empirical models (Case studies: Zahedan and Bojnurd stations). *J Atmos Sol Terr Phys* 2015;123:39–47.

6. Khorasanizadeh H, Mohammadi K. Prediction of daily global solar radiation by day of the year in four cities located in the sunny regions of Iran. *Energy Convers Manag* 2013;76:385–392.

7. Khorasanizadeh H, Mohammadi K, Jalilvand M. A statistical comparative study to demonstrate the merit of day of the year-based models for estimation of horizontal global solar radiation. *Energy Convers Manag* 2014;87:37–47.

8. Behrang MA, Assareh E, Ghanbarzadeh A, Noghrehabadi AR. The potential of different artificial neural network (ANN) techniques in daily global solar radiation modeling based on meteorological data. *Solar Energy* 2010;84:1468–1480.

9. M. Kasra, Shamshirband S, Anisi MH, Alam KA, Petković D. Support vector regression based prediction of global solar radiation on a horizontal surface. *Energy Convers Manag* 2015;91:433–41.

10. Li M-F, Liu H-B, Guo P-T, Wu W. Estimation of daily solar radiation from routinely observed meteorological data in Chongqing, China. *Energy Convers Manag* 2010;51:2575–2579.

11. Zang H, Xu Q, Bian H. Generation of typical solar radiation data for different climates of China. *Energy* 2012;38:236–248.

12. Zhao N, Zeng X, Han S. Solar radiation estimation using sunshine hour and air pollution index in China. *Energy Convers Manag* 2013;76:846–851.

13. Li H, Ma W, Lian Y, Wang X. Estimating daily global solar radiation by day of year in China. *Appl Energy* 2010;87:3011–3017.

14. Piri J, Shamshirband S, Petković D, Tong CW. Rehman MH. prediction of the solar radiation on the earth using support vector regression technique. *Infrared Phys Technol* 2015;68:179–185.

15. Park J-K, Das A, Park J-H. A new approach to estimate the spatial distribution of solar radiation using topographic factor and sunshine duration in South Korea. *Energy Convers Manag* 2015;101: 30–39.

16. Manzano A, Martín ML, Valero F, Armenta C. A single method to estimate the daily global solar radiation from monthly data. *Atmos Res* 2015;166:70–82.

17. Yang K, Koike T, Ye B. Improving estimation of hourly, daily, and monthly solar radiation by importing global data sets. *Agric Meteor* 2006;137:43–55.

18. Tang W, Yang K, He J, Qin J. Quality control and estimation of global solar radiation in China. *Sol Energy* 2010;84:466–475.

19. Teke A, Yýldýrým HB. Estimating the monthly global solar radiation for Eastern Mediterranean Region. *Energy Convers Manag* 2014;87: 628–635.

20. Li H, Ma W, Lian Y, Wang X, Zhao L. Global solar radiation estimation with sunshine duration in Tibet, China. *Renew Energy* 2011;36: 3141–3145.

21. Liu J, Liu J, Linderholm HW, Chen D, Yu Q, Wu D, et al. Observation and calculation of the solar radiation on the Tibetan Plateau. *Energy Convers Manag* 2012;57:23–32.

22. Katiyar AK, Pandey CK. Simple correlation for estimating the global solar radiation on horizontal surfaces in India. *Energy* 2010;35: 5043–5048.

23. Wan Kevin KW, Tang HL, Yang L, Lam JC. An analysis of thermal and solar zone radiation models using an Angstrom–Prescott equation and artificial neural networks. *Energy* 2008;33:1115–1127.

24. Lam JC, Wan KKW, Yang L. Solar radiation modelling using ANNs for different climates in China. *Energy Convers Manag* 2008;49: 1080–1090.

25. Landeras G, Lopes JJ, Kýsý O, Shiri J. Temperature based daily incoming solar radiation modeling based on gene expression programming, neuro-fuzzy and neural network computing techniques. *Geophysical Research Abstracts* vol. 14, EGU2012–2720, 2012 EGU General Assembly 2012.

26. Chelbi M, Gagnon Y, Waewsak J. Solar radiation mapping using sunshine duration-based models and interpolation techniques: application to Tunisia. *Energy Convers Manag* 2015;101:203–215.

27. Chen J-L, Li G-S, Wu S-J. Assessing the potential of support vector machine for estimating daily solar radiation using sunshine duration. *Energy Convers Manag* 2013;75:311–318.

28. Zhou J, Yezheng W, Gang Y. General formula for estimation of monthly average daily global solar radiation in China. *Energy Convers Manag* 2005;46:257–268.

29. Sun H, Yan D, Zhao N, Zhou J. Empirical investigation on modeling solar radiation series with ARMA–GARCH models. *Energy Convers Manag* 2015;92:385–395.

30. Hacer D, Aydin H. Sunshine-based estimation of global solar radiation on horizontal surface at Lake Van region (Turkey). *Energy Convers Manag* 2012;58:35–46.

31. Ayodele TR, Ogunjuyigbe ASO. Prediction of monthly average global solar radiation based on statistical distribution of clearness index. *Energy* 2015;90:1733–1742.

32. Khorasanizadeh H, Mohammadi K. Introducing the best model for predicting the monthly mean global solar radiation over six major cities of Iran. *Energy* 2013;51:257–266.

33. Yao W, Li Z, Xiu T, Lu Y, Li X. New decomposition models to estimate hourly global solar radiation from the daily value. *Sol Energy* 2015;120:87–99.

34. Yao W, Li Z, Wang Y, Jiang F, Hu L. Evaluation of global solar radiation models for Shanghai, China. *Energy Convers Manag* 2014;84:597–612.

35. Pan T, Wu S, Dai E, Liu Y. Estimating the daily global solar radiation spatial distribution from diurnal temperature ranges over the Tibetan Plateau in China. *Appl Energy* 2013;107:384–393.

36. Mohammadi K, Shahaboddin S, Chong WT, Muhammad A. A new hybrid support vector machine-wavelet transform approach for

estimation of horizontal global solar radiation. *Energy Convers Manag* 2015;92:162–171.

37. Antonio D, Andrisani A, Bonnici M, Graditi G, Leanza G, Madonia M, et al. A new correlation between global solar energy radiation and daily temperature variations. *Sol Energy* 2015;116:117–124.

38. El-Sebaii AA, Al-Hazmi FS, Al-Ghamdi AA, Yaghmour SJ. Global, direct and diffuse solar radiation on horizontal and tilted surfaces in Jeddah, Saudi Arabia. *Appl Energy* 2010;87:568–576.

39. El-Sebaii AA, Al-Ghamdi AA, Al-Hazmi FS, Faidah AS. Estimation of global solar radiation on horizontal surfaces in Jeddah, Saudi Arabia. *Energy Policy* 2009;37:3645–3649.

40. Jiang T. Determinant of sample correlation matrix with application. *Ann Appl Probab* 2019;29(3):1356–1397.

41. Bakirci K. Models of solar radiation with hours of bright sunshine: a review. *Renew Sustain Energy Rev* 2009;13:2580–2588.

42. Janjai S, Pankaew P, Laksanaboonsong J. A model for calculating hourly global solar radiation from satellite data in the tropics. *Appl Energy* 2009;86:1450–1457.

43. Wan Nik WB, Ibrahim MZ, Samo KB, Muzathik AM. Monthly mean hourly global solar radiation estimation. *Sol Energy* 2012;86: 379–387.

44. Adaramola MS. Estimating global solar radiation using common meteorological data in Akure, Nigeria. *Renew Energy* 2012;47:38–44.

45. Ozgoren M, Bilgili M, Sahin B. Estimation of global solar radiation using ANN over Turkey. *Expert Syst Appl* 2012;39:5043–5051.

46. Senkal O. Modeling of solar radiation using remote sensing and artificial neural network in Turkey. *Energy* 2010;35:4795–4801.

47. Lanre O, Mekhilef S, Shamshirband S, Petković D. Adaptive neuro-fuzzy approach for solar radiation prediction in Nigeria. *Renew Sustain Energy Rev* 2015;51:1784–1791.

48. Chen RS, Lu SH, Kang ES, et al. Estimating daily global radiation using two types of revised models in China. *Energy Convers Manage* 2006;47:865–878.

49. Yorukoglu M, Celik AN. A critical review on the estimation of daily global solar radiation from sunshine duration. *Energy Convers Manage* 2006;47:2441–2450.

50. Dogniaux R, Lemoine M. Classification of radiation sites in terms of different indices of atmospheric transparency. *Solar Energy*

Research and Development in the European Community, Series F, vol. 2. Dordrecht, Holland: Reidel, 1983.

51. Bahel V, BakhshH, Srinivasan, R. Correlation for estimation of global solar radiation. *Energy* 1987;12:131–135.

52. Jain S, Jain PC. A comparison of the Angstrom-type correlations and the estimation of monthly average daily global irradiation, *Solar Energy* 1988;40(2):93–98.

53. Sfetsos A, Coonik AH. Univariate and multivariate forecasting of hourly solar radiation with artificial intelligence techniques. *Solar Energy* 2000;68(2):169–178.

54. Derse MS. Batman'ýn Ýklim Koþullarýnda Eðimli Düzleme Gelen Güneþ Iþýnýmýnýn Farklý Açý Deðerlerinde Belirlenmesi, 2014, pp. 37–47.

55. Miguel AD, Bilbao J, Aguiar R, Kambezidis H, Negro E. Diffuse solar irradiation model evaluation in the North Mediterranean Belt area. *Solar Energy* 2001;70:143–153.

56. Notton G, Poggi P, Cristofari C. Predicting hourly solar irradiations on inclined surfaces based on the horizontal measurements: performances of the association of well-known mathematical models. *Energy Convers Manag* 2006;47:1816–1829.

57. Erbs DG, Klein SA, Duffie JA. Estimation of the diffuse radiation fraction for hourly, daily and monthly-average global radiation. *Solar Energy* 1982;28(4):293–302.

Chapter 4

Improving Heat Transfer Efficiency with Innovative Turbulence Generators

Yang Yang,[a] David S.-K. Ting,[a] and Steve Ray[b]

[a]Turbulence and Energy Laboratory, University of Windsor, Ontario, Canada
[b]Essex Energy, Oldcastle, Ontario, Canada

yang19n@uwindsor.ca

Enhancing heat transfer can improve the efficiency of the underlying heat exchangers of many engineering systems and thus, in many applications, assist in mitigating climate change. Among the numerous innovative techniques, a passive turbulence generator is a simple and potent means that can be exploited to boost energy conversion efficiency of solar thermal, photovoltaic, and other renewable energy systems. This chapter presents a review of passive turbulence generators for convection heat transfer augmentation, including grids, ribs, winglets, wings, and flexible turbulence generators. It is found that winglets and wings can produce long-lasting longitudinal vortices, which promote large-scale mixing, disrupt the boundary layer, and increase the turbulence intensity over an extended region, effectively boosting the heat transfer rate. Furthermore, flexible turbulence generators also provide significant potential in furthering the heat transfer augmentation by oscillating one or more turbulent vortex

Climate Change and Pragmatic Engineering Mitigation
Edited by Jacqueline A. Stagner and David S.-K. Ting
Copyright © 2022 Jenny Stanford Publishing Pte. Ltd.
ISBN 978-981-4877-97-8 (Hardcover), 978-1-003-25658-8 (eBook)
www.jennystanford.com

streets between the surface and the freestream. In other words, marrying long-lasting vortex streets with periodic fluttering seems promising in maximizing heat convection.

4.1 Introduction

A straightforward way to mitigate climate change is to improve the efficiency of engineering systems. Every bit of the performance improvement is translated into less resource usage, reduced entropy generation, decreased emissions, and ultimately less adverse effect on our environment. Many engineered systems involve effective heat transfer to function. They typically involve heat exchangers in one form or another. Renewable energy systems such as a solar photovoltaic (PV) and solar thermal system are no exception. Currently, the prevailing commercial silicon PV modules have an energy conversion efficiency of around 16% [1], and this decreases under abundant solar energy exposure as the cell temperature increases. On average, the energy conversion efficiency of a silicon module decreases by 0.2% per degree Celsius increase in cell temperature [2]. This implies cooling PV panels offers a significant potential for improving their performance. By the end of 2013, PV systems generated 160 TWh/year of clean electricity, resulting in a 140 million metric tons of CO_2 reduction per year [1]. Even a 1% increase in power output, around 5°C decrease induced by augmented heat transfer, could contribute to more than 1 million metric tons of CO_2 reduction per year. This can definitely play an important role in mitigating climate change.

4.2 Techniques of Heat Transfer Enhancement

4.2.1 Active Methods

Heat transfer enhancement can be categorized into two groups: active and passive cooling methods. The active approaches require external power input to realize the heat transfer enhancement. Some examples include fluid stream [3], jet impingement, vibration, etc. [4].

An innovative technique using water spray was proposed by Nizetic et al. [5]. The solar PV panel was studied on a clear summer day with surrounding temperature ranging from 27 to 30°C. Without any cooling, the average panel temperature was 56°C, and the maximum power output was 35 W. When they simultaneously sprayed water on the front and back sides of the PV panel, Fig. 4.1, the average panel temperature was reduced to 24°C. The corresponding maximal power output increased by 16%, to 41 W.

Figure 4.1 The front and backside of the PV panel with water spray system [5].

Kianifard et al. [6] conducted an investigation on a novel design of a water-cooled PV panel, see Fig. 4.2. A serpentine half pipe was employed on the backside of the panel instead of the full circle pipe. According to their study, the new design provided a 13% increase in the thermal efficiency and a 0.6% increase in the electrical efficiency, compared to the conventional circular design.

Figure 4.2 Experimental PV panel and the back of the panel [6].

Chen et al. [7] investigated the thermal performance of a building-integrated photovoltaic-thermal (BIPV/T) system. Via a variable speed fan, the outdoor air was drawn and transferred beneath the PV panels. It was found that the air flow had a significant effect on the PV. The temperature of the PV was around 50 to 60°C, while the roof without BIPV arrays, i.e., no air flow underneath, was higher than 70°C. In addition, the heated air from the PV arrays could also be utilized for the domestic hot water, improving the overall thermal performance.

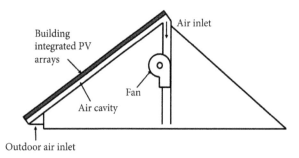

Figure 4.3 Schematic of building-integrated photovoltaic-thermal system; redrawn based on Ref [7].

Rajaseenivasan et al. [8] developed an experimental study, examining the performance of an impinging jet solar air heater. In this study, the nozzle diameter of the jet was varied from 3 to 7 mm, and the mass flow rate was altered from 0.012 to 0.016 kg/s. Also, attack angles of 0°, 10°, 20°, 30°, 60°, and 90° were considered. A maximum thermal enhancement factor of around 2.2 was obtained for the 5 mm diameter and 30° attack angle nozzle with a mass flow rate of 0.016 kg/s.

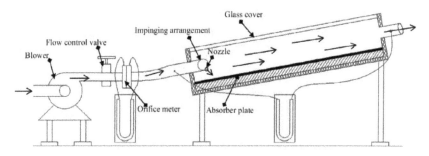

Figure 4.4 Schematic of an impinging jet solar air heater [8].

Hsu et al. [9] explored the heat transfer characteristics of a pulsed jet impinging on a flat plate; see Fig. 4.5. They found that the convective heat transfer increases with increasing convective velocity of the vortex ring and also the fluctuation intensity. The average Nusselt number enhancement was up to 100% over the range of studied conditions.

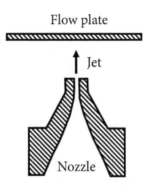

Figure 4.5 Experimental schematic of a pulsed jet impinging on a flat plate; redrawn based on Ref [9].

An experimental and numerical study on heat transfer augmentation via ultrasonic vibration in a double-pipe heat exchanger was developed by Setareh et al. [10]; see Fig. 4.6. The effect of the transmitted acoustic power on the overall heat transfer coefficient was investigated. Under the condition of Q_h = 0.5 L/min and Q_c = 1 L/min, where Q_h and Q_c are the volume flow rate of hot and cold fluid, respectively, the 120 W-transmitted acoustic power could induce approximately 57% overall heat transfer coefficient augmentation.

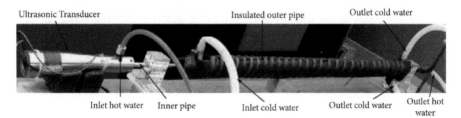

Figure 4.6 Experimental setup of double-pipe heat exchanger with ultrasonic transducer [10].

An innovative design of active vacuum system was proposed by Roesle et al. [11] to minimize the heat loss from a parabolic trough absorber tube, and thus, enhance the overall thermal efficiency. They found that the heat loss was approximately 400 Wm^{-2} at a temperature of 400°C, less than 10% of the energy gain. In a similar study, Daniel et al. [12] conducted a numerical study, comparing the thermal performance between the evacuated parabolic trough receiver and the non-evacuated receiver. And they found that the heat loss for the evacuated receiver was around 100 Wm^{-2}, much less than that associated with the non-evacuated receiver, which was approximately 500 Wm^{-2}.

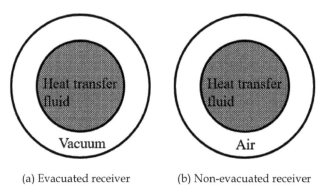

(a) Evacuated receiver   (b) Non-evacuated receiver

Figure 4.7 Schematic of (a) evacuated receiver and (b) non-evacuated receiver; redrawn based on Ref [12].

Park and Gharib [13] experimentally investigated the heat convection from a cylinder. With an oscillating mechanism, the cylinder was forced to vibrate in a sinusoidal motion with frequency of 0 to 14 Hz, corresponding to a Strouhal number ranged from 0 to 1 based on the cylinder diameter and freestream velocity. According to their results, a peak value of heat convection was observed near a Strouhal number of around 0.21, corresponding to the synchronization of mechanical oscillations with the vortex shedding.

4.2.2   Passive Methods

As mentioned above, a significant drawback of the active cooling approach is the necessity of external power. Other than higher initial costs, more maintenance is also required because of the added complexity. For this reason, the simple passive approach is often preferred. Among others, employing an appropriate turbulence generator is a promising and simple approach for improving heat convection. Accordingly, it has drawn significant attention in recent years. We will focus on the passive methods for the remaining of the chapter.

4.3 Turbulence Generator

4.3.1 Rigid Type Turbulence Generator

Turbulence generators including grid, rib, cylinder, winglet, and wing have been proposed and studied in an effort to boost the heat transfer rate. In this section, the rigid type is introduced first, and the flexible ones follow thereafter.

4.3.1.1 Grid

Grids are widely employed in many engineering applications, for reducing low frequency noise [14], improving the turbulent combustion rate [15], and augmenting the heat transfer. An experimental study of heat transfer enhancement of a cylinder via different types of grids was developed by Melina et al. [16]. In this study, a regular square-mesh bi-planar grid, a fractal-square grid, and a single-square grid were examined, the blockage ratio of the grids were 32%, 25%, and 20%, respectively. According to their results, the fractal square grid showed superior performance, which provided a higher heat transfer rate from the cylinder in the turbulence production region than that of the single square grid, despite the turbulence intensity being lower. At a larger downstream distance in the decay region, the heat transfer is similar for the fractal square grid and the single square grid, and both are higher than that of the regular square-mesh bi-planar grid. In their other study [17], eleven different grids, including the regular grid, the fractal square grid, and the single-square grid used in the previous study, were investigated for their effectiveness in heat transfer enhancement of a flat plate. A new type of grid, multi-scale inhomogeneous grids, were also considered, noting that all the grids have the same blockage ratio of 28%. Their results demonstrated that the multi-scale inhomogeneous grid led to the most effective overall heat transfer rate, furthering the heat transfer performance by around 20% in the near-grid region, compared to the no-grid case. However, due to the dissipative nature of turbulence, the heat transfer augmentation dropped to only 10% farther downstream. Based on curve-fittings, the effect of the turbulence intensity on the flat plate heat transfer was identified, and a relation of $St/St_\infty = 0.02\ u'/U$ was concluded, where St/St_∞ denotes the convective heat transfer enhancement and u'/U is the streamwise turbulence intensity.

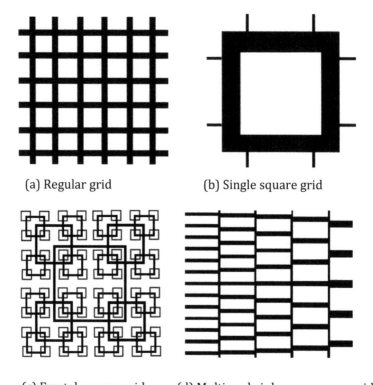

(a) Regular grid   (b) Single square grid

(c) Fractal-square grid   (d) Multi-scale inhomogeneous grid

Figure 4.8 Sketches of turbulence-generating grids: (a) regular grid; (b) single-square grid; (c) fractal-square grid; (d) multi-scale inhomogeneous grid; redrawn based on Refs [16, 17].

4.3.1.2 Rib and cylinder

To further heat transfer rate, larger organized vortical flow structures are considered in addition to the rapidly dissipating turbulence. As these secondary swirling structures move across the heated surface, they enhance the mixing of the cooler freestream with the hotter fluid near the heated surface, further enhancing the heat convection. These vortices could be divided into two main categories: transverse and longitudinal vortex, where the rotation axis is closely perpendicular to and aligned with, respectively, the freestream direction. Among others, ribs and cylinders are common transverse vortex generators that are widely developed and utilized in many engineering applications due to their structural simplicity. A summary of studies using ribs are presented in Table 4.1. The studies of grid are also included in the table for comparison. It is interesting to note that the introduction of organized vortical flow structures seems to have a significant impact on the heat transfer, boosting the maximum relative heat transfer enhancement from 20%, induced by an inhomogeneous grid [17], to as high as 120%, provided by the crescent rib [21].

Yang et al. [18] investigated the influence of a square cylinder attack angle on heat transfer enhancement from a surrogate PV panel; see Fig. 4.9. The attack angle from 15° to 75° with 15° increment was scrutinized at a wind velocity of 5 m/s, a Reynolds number, based on the width of the square wire, of around 1300. They found that the cylinder with a 60° attack angle provided the highest heat transfer augmentation, with a maximum normalized Nusselt number of around 1.8, due to it generating the strongest turbulence intensity. This equates a 2.8% gain in the harnessed power output.

Changcharoen and Eiamsa-ard [19] conducted a numerical investigation of heat transfer in a channel with attached/detached rib arrays; see Fig. 4.10. They found that the detached-clearance ratio of 0 (attached rib), 0.1, 0.2, 0.3, and 0.4 were able to improve the average heat transfer rates by around 60%, 75%, 71%, 64%, and 56%, respectively. In addition to the turbulence intensity, a significant correlation between the flow recirculation (transverse vortex) and the heat transfer was found. With the detached-clearance ratio increased from 0.1 to 0.4, a decreasing trend of the heat transfer enhancement was clear, which was observed to be directly related to the reduction of the recirculation size.

Table 4.1 Summary of rigid type grids and ribs

		Classification		Remarks	
Authors	Type	Advantages	Studied case	Best condition for heat transfer	Relative maximal heat transfer enhancement (compared to the reference case)
Melina et al. [16, 17]	Grid	Turbulence is induced, enhancing heat and mass transporting	Regular square-mesh bi-planar grid, fractal-square grid, a single-square grid, and multi-scale inhomo-geneous grids	Multi-scale inhomo-geneous grids	20%
Yang et al. [18]	Rib	Transverse vortex structures are induced in addition to the dissipative turbulence, furthering heat transfer enhancement	Attack angle from 15° to 75°	60°	80%
Changcharoen and Eiamsa-ard [19]	Rib		Detached-clearance ratio from 0 to 0.4	0.1	75%
Tanda [20]	Rib		Pitch-to-height ratios, 6.66, 10, 13.33, and 20	13.33	45%
Xie et al. [21]	Rib		Straight rib and the crescent rib	The crescent rib concave to the streamwise direction (longi-tudinal vortex occurs)	120%

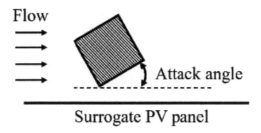

Figure 4.9 Attack angle of square cylinder; redrawn based on Ref [18].

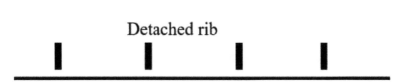

Figure 4.10 Detached rib in the channel; redrawn based on Ref [19].

Tanda [20] experimentally explored heat convection in a rectangular channel with an angled rib turbulator; see Fig. 4.11. Four different pitch-to-height ratios, 6.66, 10, 13.33, and 20, were investigated. Among the studied conditions, the best heat transfer performance was obtained at a pitch-to-height ratio of 13.33 for the one-ribbed wall channel, providing a heat transfer augmentation as high as 1.45. For the two-ribbed wall channel, the optimal pitch-to-height ratio was found to be 10, which induced a maximal heat transfer enhancement of around 1.2.

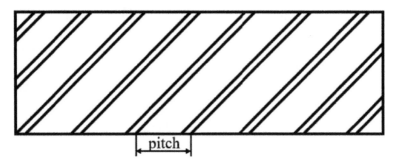

Figure 4.11 Ribbed surface; redrawn based on Ref [20].

In addition to the conventional straight square rib, some new types of ribs have also been proposed for furthering the heat transfer rate. Xie et al. [21] explored the turbulent flow characteristics and the heat transfer enhancement in a channel with crescent ribs; see Fig. 4.12. The straight rib, the crescent rib concave to the streamwise direction, and the crescent rib convex to the streamwise direction were investigated. The crescent rib was able to generate longitudinal vortices, intensifying the flow mixing, increasing the turbulent kinetic energy, and reducing the boundary layer thickness. In summary, the crescent rib concave to the streamwise direction provided the best thermal performance, providing a normalized Nusselt number as high as 2.2, around 21–40% higher than that of the straight ribbed channel, along with an inevitably higher pressure drop of around 15–80%.

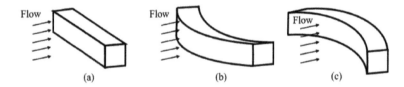

Figure 4.12 Schematic of (a) straight rib, (b) crescent rib concave to the streamwise direction, and (c) crescent rib convex to the streamwise direction; redrawn based on Ref [21].

4.3.1.3 Winglet and wing

As mentioned in Ref [21], longitudinal vortices tend to enhance heat transfer more effectively. These longitudinal vortical streets can drastically outlast the turbulence created by the more conventional turbulence generators such as grids and ribs. Among many others, winglets and wings, such as those shown in Fig. 4.13, are commonly used as longitudinal vortex promoters for the heat transfer enhancement [22]. Table 4.2 presents a brief summary of winglets and wings. Discussion of each of these studies will follow accordingly.

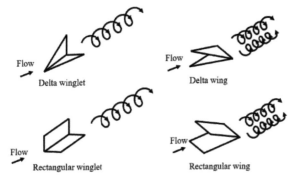

Figure 4.13 Winglets and wings; redrawn based on Ref [22].

Table 4.2 Summary of rigid type winglets and wings

	Classification			Remarks	
Authors	**Type**	**Advantages**	**Studied case**	**Best condition for heat transfer**	**Relative maximal heat transfer enhancement (compared to the reference case if there is no specific mention)**
Wu et al. [23]	Winglet	More effective and long-lasting longitudinal vortices are generated, which can drastically outlast the turbulence over the whole heated surface	Attack angle from 30° to 60°	60°	100%
Abdollahi and Shams [24]	Winglet		Attack angle from 15° to 90°	45°	8% higher than that at 15°
Khanjian et al. [25]	Winglet		Roll angle from 20° to 90°	90°	60%
Li et al. [26]	Winglet		Arrangements	Common-flow-up arrangement with vertical edge being the trailing edge	22%
Promvonge et al. [27]	Winglet		Relative heights RB from 0.1 to 0.2 Relative pitches RP from 0.5 to 2	RB = 0.2 and RP = 0.5	380%

(Continued)

Table 4.2 (*Continued*)

	Classification			Remarks	
Authors	Type	Advantages	Studied case	Best condition for heat transfer	Relative maximal heat transfer enhancement (compared to the reference case if there is no specific mention)
Pourhedayat et al. [28]	Winglet		Transversal distance from 0 to 40 mm, tube diameter = 47 mm	20 mm	140%
Gholami et al. [29]	Winglet	More effective and long-lasting longitudinal vortices are generated, which can drastically outlast the turbulence over the whole heated surface	Wavy and flat winglet	Wavy winglet	58%
Oneissi et al. [30, 31]	Winglet		Inclined projected winglet with protrusion and conventional delta winglet	Inclined projected winglet with protrusion	7.1% compared to the conventional delta winglet
Haik et al. [32]	Winglet		Curved and flat winglets	Concave-curved winglet with 75° arc angle	22%
Biswas et al. [33]	Winglet and Wing		Winglet and Wing	Wing	34%
Sheikhzadeh et al. [34]	Wing		The rectangular, triangle and trapezoid wing	Rectangular wing	13% and 30% higher than those of trapezoidal and triangular wings

Wu et al. [23] experimentally investigated the effect of delta winglet attack angle on the heat transfer from a flat surface as depicted in Fig. 4.14. The single delta winglet was positioned on a heated surface, under a Reynolds number based on winglet height of 6000. The attack angle of winglet was 30°, 45°, and 60° with respect to the streamwise direction. According to their results, a heat transfer augmentation region was obtained downstream of the winglet, due to the scooping motions of the longitudinal vortices, turbulence fluctuation and the thinning of boundary layer. With attack angle increases from 30° to 60°, the maximum heat transfer enhancement increases from around 1.5 to 2.0. They pointed out that, in addition to the turbulent kinetic energy, the secondary flow motion also significantly influences the heat transfer enhancement -- the larger velocity toward the heated surface, the better heat transfer performance. The near-surface streamwise velocity was found to have a moderate effect, while the influence of the boundary layer thickness was marginal.

Figure 4.14 Attack angle of delta winglet; redrawn based on Ref [23].

The optimization of shape and attack angle of the winglet was developed by Abdollahi and Shams [24]. Rectangular, trapezoidal, and delta winglets were employed in this research, with attack angle of 15°, 30°, 45°, 60°, and 90°. The Reynolds number based on the hydraulic diameter of the channel was varied from 117 to 467. Their results showed that the rectangular winglet provided the best heat transfer augmentation, followed by the trapezoidal winglet and then, delta winglet. Further, when the Reynolds is low, the heat transfer was observed to be insensitive to the attack angle. But at a higher Reynolds number of 467, the attack angle showed a significant effect, with average Nusselt number of around 13.5 at 45°, approximately 8% higher than that at 15°. It was also mentioned in the study that, the longitudinal vortical motions induced by the winglets resulted in a higher fluid mixing between the cold freestream fluid and the hot fluid near the heated surface, and thus, significantly increased the heat transfer rate.

The effect of roll angles of rectangular winglets, ranging from 20° to 90°, was investigated by Khanjian et al. [25]. The attack angle in this study was maintained at 30°, and the Reynolds number based on the channel hydraulic diameter of 456 and 911 were selected. They concluded that the heat transfer improved with the increase of the roll angle, but so does the pressure drop. The global Nusselt number of around 11 was obtained at 20° roll angle and a Reynolds number of 911. The Nusselt number increased to approximately 16 when the roll angle increased to 90°, due to the stronger vortices, with a 12 times higher helicity peak value than that of 20° roll-angle winglet. If the pressure drop was also considered, the overall thermal enhancement factor reached a maximum value of 1.32 for a 70° roll angle.

Also worth mentioning is a numerical investigation by Li et al. [26]. Four different arrangements of the delta winglet pairs in a jacket were explored, including common-flow-down with vertical edge being the trailing edge (CFD-V), common-flow-up with vertical edge being the trailing edge (CFU-V), common-flow-down with inclined edge being the trailing edge (CFD-I) and common-flow-up with inclined edge being the trailing edge (CFU-I); see Fig. 4.15. According to their results, the common-flow-up configuration provided better heat transfer than that of the common-flow-down arrangement for Reynolds number based on pipe diameter from 4000 to 18000. Overall, the CFU-V had the superior heat transfer performance due to the stronger, larger-scale, and slower-dissipative vortices, providing an average Nusselt number enhancement up to approximately 22%, compared to the smooth jacket in the absence of winglets.

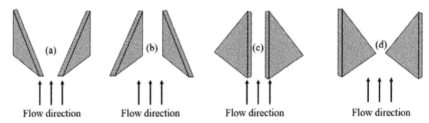

Figure 4.15 The arrangement of the delta winglet pair: (a) Common-flow-down with vertical edge being the trailing edge. (b) Common-flow-up with vertical edge being the trailing edge. (c) Common-flow-down with inclined edge being the trailing edge. (d) Common-flow-up with inclined edge being the trailing edge; redrawn based on Ref [26].

Promvonge et al. [27] conducted an experimental and numerical investigation on heat transfer enhancement in a tube with rectangular winglet, as portrayed in Fig. 4.16, aiming at optimizing the height, pitch, and arrangement of the winglet. The Reynolds number was varied from 4200 to 25800. Two arrangements of winglets are introduced, i.e., V-tip pointing upstream and downstream, with three relative heights RB (winglet height/tube diameter = 0.1, 0.15 and 0.2) and four different relative pitches RP (longitudinal pitch of winglets/tube diameter = 0.5, 1.0, 1.5 and 2.0). They concluded that the larger height and smaller pitch resulted in a higher Nusselt number as well as friction factor. The winglet with RB = 0.2 and RP = 0.5 increased the maximum Nusselt number and friction factor by about 3.8 times and 18.8 times, respectively. The best overall thermal performance enhancement factor was obtained at RB = 0.15 and RP = 1.0, up to 1.99 and 2.02 for the V-tip pointing upstream and downstream, respectively. They also postulated that the key reason for the heat transfer enhancement was the impinging jets on the heated surface induced by the secondary flow motion.

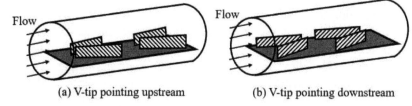

(a) V-tip pointing upstream   (b) V-tip pointing downstream

Figure 4.16 Winglet arrangement: (a) V-tip pointing upstream; (b) V-tip pointing downstream; redrawn based on Ref [27].

In another numerical study, a novel arrangement of the delta winglet was proposed by Pourhedayat et al. [28]. In this study, the winglets were placed on both sides of the plate, Fig. 4.17, furthering the heat transfer augmentation through a 47 mm-diameter tube. The backward and forward configurations were analyzed in this study. In addition, the transversal distance of the winglets was also investigated. Their results pointed out that the forward configuration resulted in a higher heat transfer rate than that of the backward configuration. Moreover, smaller longitudinal pitch provided a stronger heat transfer rate, which is in agreement with Ref [27]. Concerning the effect of transversal pitch, it is found that the winglet should not be placed very close to the wall or the center of the tube. The winglet with a transversal distance of 20 mm induced the best heat transfer performance. This was attributed to the best fluid mixing condition in the cross-section of the tube, providing a heat transfer enhancement of around 2.4, compared to the reference case without the winglet at a Reynolds number of 12000.

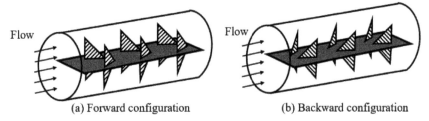

(a) Forward configuration   (b) Backward configuration

Figure 4.17 Partial view of the delta winglet in the tube: (a) forward configuration; (b) backward configuration; redrawn based on Ref [28].

Besides the conventional delta and rectangular winglet, some additional innovative winglet designs were also developed to further enhance the heat convection. For example, Gholami et al. [29] explored the effect of wavy rectangular winglet on the heat transfer of a fin-and-tube compact heat exchanger. The winglets were positioned behind the tube, with an attack angle of 30°, see Fig. 4.18, for a Reynolds number ranging from 400 to 800. The wavy-down winglet, wavy-up winglet, convectional flat rectangular winglet, and the reference case without any winglet were investigated. They concluded that the wavy-down winglet and the wavy-up winglet could further decrease the stationary wake region behind the tube compared to the flat winglet, and thus, a better Nusselt number could be obtained. Among their studied cases, the wavy-up winglet provided the most effective heat transfer, with Nusselt number of around 15.0. And the Nusselt number for wavy-down winglet, flat winglet, and the reference case are around 14.5, 13.5, and 9.5, respectively.

Oneissi et al. [30] proposed a different winglet, calling it an inclined projected winglet. The inclined angle, shown in Fig. 4.19, was varied from 25° to 60° to investigate its effect on the heat transfer enhancement and the pressure drop in a channel. The Reynolds number was varied from 270 to 30000. It was found that the delta winglet with 30°~35° inclined angle had a superior overall heat transfer performance, providing a small increase of 3.1% in average Nusselt number and a significant decrease of 55% in friction factor, compared to the conventional delta winglet. This has been attributed to the 35.5% higher intensification of vortex. In their other study [31], the inclined projected winglet with protrusions was explored, Fig. 4.20, aiming at furthering the heat transfer enhancement. The 30° inclined projected winglet was employed in this study, and the semi-sphere protrusion was located downstream of the winglet. According to their results, the inclined projected winglet could provide an overall thermal enhancement of around 4.8% and the protrusion could further boost the heat transfer improvement to 7.1%, compared to the conventional delta winglet. The further enhancement was because the protrusions were able to generate two additional small vortex structures near the wall.

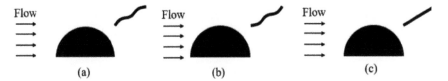

Figure 4.18 Winglets in the heat exchanger: (a) wavy-down winglet; (b) wavy-up winglet; (c) convectional flat winglet; redrawn based on Ref [29].

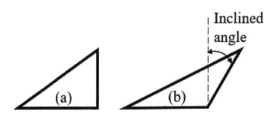

Figure 4.19 Schematic of winglet: (a) conventional delta winglet; (b) inclined delta winglet; redrawn based on Ref [30].

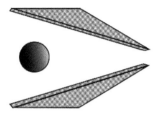

Figure 4.20 Schematic of winglet with protrusion; redrawn based on Ref [31].

Haik et al. [32] conducted a numerical study, demonstrating the heat transfer enhancement in a channel with the curved rectangular winglet. The concavely and convexly curved winglets, see Fig. 4.21, with arc angle ranging from 15° to 75°, were investigated at a Reynolds number, based on the height of the channel, of 3000. Their results showed that the concave-curved winglet induced more disruption in boundary layer growth and fluid mixing, due to the wider horseshoe vortices and the higher strength of longitudinal vortices. These resulted in higher heat transfer rate as well as higher pressure loss. In summary, the maximum heat transfer rate was obtained when using the concave-curved winglet with a 75° arc angle, about 3% and 22% higher than those of the flat winglet case and the reference case in absence of the winglet, respectively.

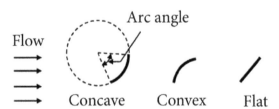

Figure 4.21 Schematic of the curve winglet; redrawn based on Ref [32].

Compared to the winglet type turbulence generators, the wing type seemed to provide a larger heat transfer enhancement, together with a higher pressure drop penalty. Biswas et al. [33] studied the flow and heat transfer characteristics in a rectangular channel with a delta wing and a pair of delta winglets. They found that, both wing and winglet could induce long-lasting longitudinal vortices. The fluid was swirling as it moved downstream, promoting the mixing of the cooler stream of the vortex core with the hotter fluid near the channel wall, and thus, augmenting the heat transfer rate. In addition, the delta wing generated stronger vortices and therefore provided a better heat transfer improvement compared to the winglet pair. The Nusselt number and friction coefficient at the exit of the channel with the delta wing was around 34% and 79% more than that of the reference clean channel, respectively. The corresponding values for the winglet pair were around 14% and 65%, respectively.

In another numerical study, the effects of the shape of the wings on heat transfer and friction factor were investigated at a Reynolds number of 200 to 1600 by Sheikhzadeh et al. [34]. Three types of wings, rectangular, triangular, and trapezoidal wings as sketched in Fig. 4.22, were studied. From the perspective of heat transfer enhancement, the rectangular wings seemed to have a better effect compared to the triangle and the trapezoid wing. The average Nusselt number induced by the rectangular wings was around 130 at a Reynolds number of 1600, approximately 13% and 30% higher than those of trapezoidal and triangular wings. The rectangular wings also induced a higher friction factor of around 0.8 at a Reynolds number of 1600, followed by the trapezoidal and the triangular wings, which was 0.5 and 0.3, respectively.

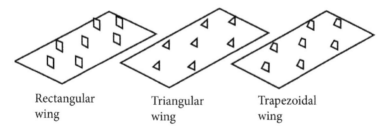

Rectangular wing   Triangular wing   Trapezoidal wing

Figure 4.22 Schematic of rectangular, triangular, and trapezoidal wing; redrawn based on Ref [34].

4.3.2 Flexible Type Turbulence Generator

Allowing the turbulence generators to vibrate when induced by the incoming flow adds another dimension, i.e., the possibility of harmonizing a portion of the swirling vortex street into furthering heat transfer augmentation. From the selected studies [35–40], listed in Table 4.3, it is clear that flexible turbulence generators could provide better heat transfer performance, compared to their rigid counterpart. Discussion of these studies will follow.

Table 4.3 Summary of flexible turbulence generators

	Classification			Remarks	
Authors	Type	Advantages	Studied case	Best condition for heat transfer	Relative maximal heat transfer enhancement (compared to the reference case if there is no specific mention)
Ali et al. [35, 36]	2D-plate	Flow induced vibration occurs, enhancing the mixing process, and thus, further improving the heat transfer augmentation	Number of flexible vortex generators (two, three, four)	Four flexible vortex generators	275%
Ali et al. [37]	Wing		Flexible and rigid wing	Flexible wing	118%
Shi et al. [38]	2D-plate		The cylinder with the flexible/rigid plate	The cylinder with the flexible plate	90%
Shi et al. [39]	2D-cylinder		Flexible/rigid cylinder	Flexible cylinder	235%
Sun et al. [40]	2D-plate		The cylinder with the flexible/rigid plate	The cylinder with the flexible plate	11% higher than that of the cylinder with a rigid plate

A numerical investigation of the heat transfer enhancement induced by a few flexible turbulence generators was conducted by Ali et al. [35]. These flexible vortex generators fixed at the channel wall, as shown in Fig. 4.23, were free to vibrate when induced by the incoming flow at a Reynolds number of 1000 and 1850. It was found that the flexible vortex generator provided a better mixture quality, a function of local concentration value, indicating the fraction of fluid from the lower part of the domain mixed in the upper part, or vice versa. As high as 98% increase of mixture quality and 134% increase of overall heat transfer rate was observed compared to their rigid counterparts. Subsequently, they conducted another numerical study, investigating the flow pattern and the heat transfer with different numbers of flexible vortex generators in a channel [36]. They found that an increase in the number of flexible vortex generators could produce a lock-in state between the flow and the vortex generators, inducing a larger motion amplitude and greatly improving fluid mixing and heat transfer. Compared to the reference empty channel, the two, three, and four flexible vortex generators could improve the overall heat transfer by 174%, 250% and 275%, respectively.

In another study by Ali et al. [37], a three-dimensional numerical investigation was executed. The flexible trapezoidal vortex generators and their rigid counterparts were employed in a circular pipe at a Reynolds number of 1500; see Fig. 4.24. They found that the vortex generator which was free to oscillate could enhance the overall heat transfer by approximately 118% compared to the reference empty pipe, while the rigid type only produced a 97% heat transfer augmentation. For the rigid case, an overheated area could be observed behind the turbulence generator, as the heat stayed trapped in the stagnant wake region and could only be transported by diffusion. The presence of the oscillation was observed to have a significant impact on the wake, eliminating the overheated region behind the turbulence generators. Instead of being trapped like that in the rigid case, the vortices in the wake region of the flexible type were convected downstream, which played an important role in the thermal energy transporting.

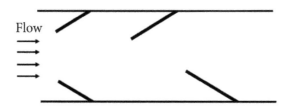

Figure 4.23 Schematic of flexible turbulence generators in the channel; redrawn based on Ref [35].

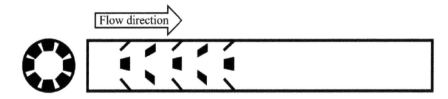

Figure 4.24 Schematic of flexible trapezoidal turbulence generators in a circular pipe; redrawn based on Ref [37].

Shi et al. [38] numerical investigated the heat transfer augmentation of a channel with vortex induced vibration. A stationary cylinder with a flexible plate was studied, compared to a cylinder with a rigid plate and a cylinder without the plate, as depicted in Fig. 4.25. The studied Reynolds number based on the diameter of the cylinder ranged from 204 to 327. Their results showed that the flow structure interaction induced by the cylinder with flexible plate could provide the stronger and closer-to-wall vortices, and thus, further disrupt the thermal boundary layer and improve the mixing process. The cylinder without the plate, and the cylinder with the rigid plate could enhance the Nusselt number by around 57% and 24%, respectively, with respect to the reference empty channel. The cylinder with the flexible plate further improved the heat transfer rate, boosting the enhancement up to 90%. In their other study, a flexible cylinder was developed for Reynolds number between 84 and 168 [39]. The vortex shedding from the cylinder induced the cylinder to vibrate periodically. This vibration caused the vortex cores to move closer to the wall and strengthened the vortex intensity. Thus, a better heat transfer augmentation was observed. Based on their study, the channel with the flexible cylinder provided an enhancement of around 235% and 51% in Nusselt number with respect to the reference clean channel and the channel with a stationary cylinder, respectively.

Sun et al. [40] conducted a numerical study on the effect of a flexible plate on the heat transfer from a circular cylinder; see Fig. 4.26. The Reynolds number was maintained at 200. The elastic modulus of the flexible plate ranged from 104 to 5×10^5. Their results showed that a large-amplitude vibration occurred when the vortex shedding frequency approached the natural frequency of the flexible plate. Thus, the 'dead water' region behind the cylinder was reduced, leading to a 11% heat transfer enhancement compared to the cylinder with a rigid plate.

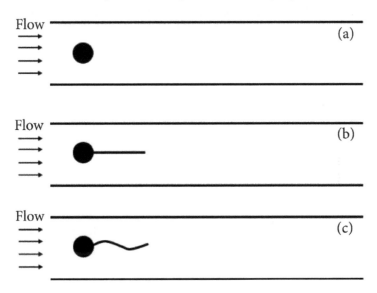

Figure 4.25 Schematic of a channel with (a) cylinder without the plate, (b) cylinder with a rigid plate, and (c) cylinder with a flexible plate; redrawn based on Ref [38].

Figure 4.26 "Dead water" region behind the (a) cylinder with rigid plate and (b) cylinder with flexible plate; redrawn based on Ref [40].

4.4 Conclusions

Augmenting heat transfer rate can improve the performance of many engineering systems and thus facilitate climate change mitigation. For example, heightening the heat transfer of a heat exchanger ultimately boosts the efficiency of the system invoking the heat exchanger. More solar energy can be harnessed by a PV panel by lowering its temperature via augmented heat convection. Among plethora of approaches for boosting heat transfer, a turbulence generator that appropriately disturbs the flow is appealing due to its simplicity. Different types of turbulence generators and their effectiveness on the heat transfer enhancement were reviewed. The following conclusions can be drawn:

1. Turbulence intensity appears to be the most important parameter in convection heat transfer enhancement.
2. Swirling flow motions that promote mixing of freestream fluid with the near-surface fluid are effective in augmenting heat transfer.
3. Vibrations of the passive turbulence generators can lead to additional heat transfer augmentation.
4. Appropriate geometrical design can be implemented to reduce stagnant fluid, such as that in the wake region behind a heat exchanger tube, and hence, augment heat transfer.
5. Marrying the long-lasting swirling flow with the fluid-induced vibration seems to provide a significant potential in maximizing the heat convection augmentation.

References

1. International Energy Agency, Technology Roadmap: Solar Photovoltaic Energy, 2014.
2. S. Kalogirou, *Solar Energy Engineering: Processes and Systems*, 2nd ed. Academic Press, UK, 2014.
3. M. Hemmat Esfe, M. H. Kamyab, and M. Valadkhani, Application of nanofluids and fluids in photovoltaic thermal system: An updated review, *Sol. Energy*, vol. 199, pp. 796–818, 2020.
4. S. Liu and M. Sakr, A comprehensive review on passive heat transfer enhancements in pipe exchangers, *Renew. Sustain. Energy Rev.*, vol. 19, pp. 64–81, 2013.
5. S. Nižetić, D. Čoko, A. Yadav, and F. Grubišić-Čabo, Water spray cooling technique applied on a photovoltaic panel: The performance response, *Energy Convers. Manag.*, vol. 108, pp. 287–296, 2016.

6. S. Kianifard, M. Zamen, and A. A. Nejad, Modeling, designing and fabrication of a novel PV/T cooling system using half pipe, *J. Clean. Prod.*, vol. 253, p. 119972, 2020.

7. Y. Chen, A. K. Athienitis, and K. Galal, Modeling, design and thermal performance of a BIPV/T system thermally coupled with a ventilated concrete slab in a low energy solar house: Part 1, BIPV/T system and house energy concept, *Solar Energy*, vol. 84, pp. 1892–1907, 2010.

8. T. Rajaseenivasan, S. R. Prasanth, M. S. Antony, and K. Srithar, Experimental investigation on the performance of an impinging jet solar heater, *Alexandria Eng. J.*, vol. 56, pp. 63–69, 2017.

9. C. M. Hsu, W. C. Jhan, and Y. Y. Chang, Flow and heat transfer characteristics of a pulsed jet impinging on a flat plate, *Heat Mass Transf. und Stoffuebertragung*, vol. 56, no. 1, pp. 143–160, 2020.

10. M. Setareh, M. Saffar-Avval, and A. Abdullah, Experimental and numerical study on heat transfer enhancement using ultrasonic vibration in a double-pipe heat exchanger, *Appl. Therm. Eng.*, vol. 159, p. 113867, 2019.

11. M. Roesle, V. Coskun, A. Steinfeld, Numerical analysis of heat loss from a parabolic trough absorber tube with active vacuum system, *J. Solar Energy Eng.*, vol. 133, p. 031015, 2011.

12. P. Daniel, Y. Joshi, A. K. Das, Numerical investigation of parabolic trough receiver performance with outer vacuum shell, *Solar Energy*, vol. 85, pp. 1910–1914, 2011.

13. H. G. Park and M. Gharib, Experimental study of heat convection from stationary and oscillating circular cylinder in cross flow, *J. Heat Transfer*, vol 123, pp. 51–62, 2001.

14. J. Nedić, B. Ganapathisubramani, J. C. Vassilicos, J. Borée, L. E. Brizzi, and A. Spohn, Aeroacoustic performance of fractal spoilers, *AIAA J.*, vol. 50, no. 12, pp. 2695–2710, 2012.

15. D. S.-K. Ting and M. D. Checkel, The importance of turbulence intensity, eddy size and flame size in spark ignited, premixed flame growth, *Proc. Inst. Mech. Eng. Part D J. Automob. Eng.*, vol. 211, no. 1, pp. 83–86, 1997.

16. G. Melina, P. J. K. Bruce, G. F. Hewitt, and J. C. Vassilicos, Heat transfer in production and decay regions of grid-generated turbulence, *Int. J. Heat Mass Transf.*, vol. 109, pp. 537–554, 2017.

17. G. Melina, P. J. K. Bruce, J. Nedić, S. Tavoularis, and J. C. Vassilicos, Heat transfer from a flat plate in inhomogeneous regions of grid-generated turbulence, *Int. J. Heat Mass Transf.*, vol. 123, pp. 1068–1086, 2018.

18. Y. Yang, A. Ahmed, D. S.-K. Ting, and S. Ray, The influence of square wire attack angle on the heat convection from a surrogate PV Panel, in *The Energy Mix for Sustaining Our Future*, Springer, pp. 103–128, 2018.

19. W. Changcharoen and S. Eiamsa-ard, Numerical investigation of turbulent heat transfer in channels with detached rib-arrays, *Heat Transf. - Asian Res.*, vol. 40, no. 5, pp. 431–447, 2011.

20. G. Tanda, Effect of rib spacing on heat transfer and friction in a rectangular channel with 45° angled rib turbulators on one/two walls, *Int. J. Heat Mass Transf.*, vol. 54, no. 5–6, pp. 1081–1090, 2011.

21. G. Xie, X. Liu, H. Yan, and J. Qin, Turbulent flow characteristics and heat transfer enhancement in a square channel with various crescent ribs on one wall, *Int. J. Heat Mass Transf.*, vol. 115, pp. 283–295, 2017.

22. L. Chai and S. A. Tassou, A review of airside heat transfer augmentation with vortex generators on heat transfer surface, *Energies*, vol. 11, p. 2737, 2018.

23. H. Wu, D. S.-K. Ting, and S. Ray, The effect of delta winglet attack angle on the heat transfer performance of a flat surface, *Int. J. Heat Mass Transf.*, vol. 120, pp. 117–126, 2018.

24. A. Abdollahi and M. Shams, Optimization of shape and angle of attack of winglet vortex generator in a rectangular channel for heat transfer enhancement, *Appl. Therm. Eng.*, vol. 81, pp. 376–387, 2015.

25. A. Khanjian, C. Habchi, S. Russeil, D. Bougeard, and T. Lemenand, Effect of rectangular winglet pair roll angle on the heat transfer enhancement in laminar channel flow, *Int. J. Therm. Sci.*, vol. 114, pp. 1–14, 2017.

26. Y. X. Li, X. Wang, J. Zhang, L. Zhang, and J. H. Wu, Comparison and analysis of the arrangement of delta winglet pair vortex generators in a half coiled jacket for heat transfer enhancement, *Int. J. Heat Mass Transf.*, vol. 129, pp. 287–298, 2019.

27. P. Promvonge, P. Promthaisong, and S. Skullong, Experimental and numerical heat transfer study of turbulent tube flow through discrete V-winglets, *Int. J. Heat Mass Transf.*, vol. 151, p. 119351, 2020.

28. S. Pourhedayat, S. M. Pesteei, H. E. Ghalinghie, M. Hashemian, and M. A. Ashraf, Thermal-exergetic behavior of triangular vortex generators through the cylindrical tubes, *Int. J. Heat Mass Transf.*, vol. 151, 2020.

29. A. A. Gholami, M. A. Wahid, and H. A. Mohammed, Heat transfer enhancement and pressure drop for fin-and-tube compact heat exchangers with wavy rectangular winglet-type vortex generators, *Int. Commun. Heat Mass Transf.*, vol. 54, pp. 132–140, 2014.

30. M. Oneissi, C. Habchi, S. Russeil, D. Bougeard, and T. Lemenand, Inclination angle optimization for 'inclined projected winglet pair' vortex generator, *J. Therm. Sci. Eng. Appl.*, vol. 11, no. 1, pp. 1–10, 2019.

31. M. Oneissi, C. Habchi, S. Russeil, T. Lemenand, and D. Bougeard, Heat transfer enhancement of inclined projected winglet pair vortex generators with protrusions, *Int. J. Therm. Sci.*, vol. 134, pp. 541–551, 2018.

32. H. Naik, S. Harikrishnan, and S. Tiwari, Numerical investigations on heat transfer characteristics of curved rectangular winglet placed in a channel, *Int. J. Therm. Sci.*, vol. 129, 2017, pp. 489–503, 2018.

33. G. Biswas, P. Deb, and S. Biswas, Generation of longitudinal streamwise vortices—a device for improving heat exchanger design, *J. Heat Transfer*, vol. 116, no. 3, pp. 588–597, 1994.

34. G. A. Sheikhzadeh, F. N. Barzoki, A. A. A. Arani, and F. Pourfattah, Wings shape effect on behavior of hybrid nanofluid inside a channel having vortex generator, *Heat Mass Transf. und Stoffuebertragung*, vol. 55, no. 7, pp. 1969–1983, 2019.

35. S. Ali, C. Habchi, S. Menanteau, T. Lemenand, and J.-L. Harion, Heat transfer and mixing enhancement by free elastic flaps oscillation, *Int. J. Heat Mass Transf.*, vol. 85, pp. 250–264, 2015.

36. S. Ali, S. Menanteau, C. Habchi, T. Lemenand, and J. L. Harion, Heat transfer and mixing enhancement by using multiple freely oscillating flexible vortex generators, *Appl. Therm. Eng.*, vol. 105, pp. 276–289, 2016.

37. S. Ali, C. Habchi, S. Menanteau, T. Lemenand, and J. L. Harion, Three-dimensional numerical study of heat transfer and mixing enhancement in a circular pipe using self-sustained oscillating flexible vorticity generators, *Chem. Eng. Sci.*, vol. 162, pp. 152–174, 2017.

38. J. Shi, J. Hu, S. R. Schafer, and C. L. Chen, Numerical study of heat transfer enhancement of channel via vortex-induced vibration, *Appl. Therm. Eng.*, vol. 70, no. 1, pp. 838–845, 2014.

39. J. Shi, J. Hu, S. R. Schafer, and C. C. L. Chen, Heat transfer enhancement of channel flow via vortex-induced vibration of flexible cylinder,

Proceedings of the ASME 2014 4th Joint US-European Fluids Engineering Division Summer Meeting, pp. 1–8, 2014.

40. X. Sun, Z. Ye, J. Li, K. Wen, and H. Tian, Forced convection heat transfer from a circular cylinder with a flexible fin, *Int. J. Heat Mass Transf.*, vol. 128, pp. 319–334, 2019.

Chapter 5

Effect of Ambient Temperature and Wind on Solar PV Efficiency in a Cold Arctic Climate

Avinash Singh, Paul Henshaw, and David S.-K. Ting

Turbulence and Energy Laboratory, University of Windsor,
Windsor, Ontario, Canada

henshaw@uwindsor.ca

A solar PV array's performance efficiency depends on the PV cell temperature, which is influenced by its radiation to the sky and convection to the ambient air, in addition to the solar irradiance. Photovoltaic arrays in the Arctic produce power at values higher than their rated capacity. For clear and sunny winter, spring, and summer days in 2017, power output data for two arrays in Iqaluit, Nunavut, Canada were analyzed. Array efficiencies were calculated based on the manufacturers' reported reference efficiencies that were adjusted by the ambient temperature and wind effects on cell temperature.

The first array performed at an estimated mean efficiency of 18.8% on a winter day, while on spring and summer days, its estimated efficiencies were 16.9% and 16.1%, respectively, compared to the PV panel reference efficiency of 15.89%. The second

Climate Change and Pragmatic Engineering Mitigation
Edited by Jacqueline A. Stagner and David S.-K. Ting
Copyright © 2022 Jenny Stanford Publishing Pte. Ltd.
ISBN 978-981-4877-97-8 (Hardcover), 978-1-003-25658-8 (eBook)
www.jennystanford.com

array efficiencies on the same winter, spring, and summer days were 19.1%, 17.0%, and 16.4%, versus the PV panel reference efficiency of 16.16%. Hence, the maximum relative efficiency enhancement was 18%. The analysis revealed that the heat loss from the panel on the winter day was mostly by radiation to the sky, while on the spring and summer days, convection to the ambient was more dominant. Considering an energy-weighted average of the efficiency enhancements for one clear and sunny day in each month, designers can expect the mean annual power output to be 4% to 7% above the rated output.

5.1   Introduction

Renewable resources include solar, hydraulic, biochemical, aeolian, and geothermal energies [1, 2]. Solar photovoltaic (PV) installation and usage over the last decade have grown tremendously because of advantages such as reducing greenhouse gas emissions, providing energy independence and sustainability, efficiency improvements, and reduction in unit cost for (PV) panels [3–5]. Accordingly, solar PV bids for large-scale electric power generation are proposing a lower levelized cost of energy than new fossil fuel power generating plants [6]. A solar PV panel/array's electrical efficiency depends on various environmental conditions such as ambient temperature, wind speed, and albedo [7, 8]. PV arrays are economical for communities where fossil fuels are very costly due to limited transportation options [9]. Many of the isolated communities located in northern Canadian territories fall into this category, as they receive the majority of their electrical energy from diesel generators. The dependence on fossil fuels results in electrical utilities with high operational and environmental costs, not to mention the hazard of storing over a year's fuel supply until the next summer's sealift. Increasing renewable energy use, such as PV technologies, supports energy sustainability and the region's future development [10].

PV technology has been perceived as a good performer in hot and dry climates due to the available solar energy throughout the year compared with cold climates having shorter days during the winter season. However, PV technology performs better in cold regions because the PV cells become cooler and thus more efficient [11–18].

This paper analyses the actual electrical efficiency performance of two separate solar PV arrays located in Iqaluit, Canada. Iqaluit is located close to the Arctic Circle, at a latitude of 63.75°N, and is the capital of the territory of Nunavut [19]. Array efficiencies were estimated from array power output data measured by the inverters and manufacturers' reference efficiencies, modified by the effects of ambient temperature and wind. No on-site measurement of solar irradiance and string-wise electrical power was available. Further analysis was performed to understand the heat loss mechanisms at various times of the year. The sensitivity of the energy balance to different equations describing the convective heat loss coefficient, sky temperature, and view factors was examined. The annual mean relative enhancement in efficiency was calculated.

5.2 Solar PV Performance in Cold and Warm Temperatures

The efficiency of a PV array increases and decreases linearly depending on the ambient temperature. As the ambient temperature increases, the array efficiency decreases, and as the ambient temperature decreases, the array efficiency increases [20]. Specifically, increasing the PV cell temperature will result in its output voltage significantly decreasing and its current slightly increasing; thus, the overall impact is a decrease in the output power [1]. Pantic et al. investigated solar PV performance in Serbia (southeastern part of Europe) during a typical winter and summer period to determine the actual PV output efficiency. During the winter period, the PV array efficiency was greater than its rated capacity by approximately 16%, while during the summer, it performed at less than rated efficiency by approximately 10% [21]. In addition, in an experiment done by Kasaeian et al. [22] where a PV array was subjected to forced convection by cold air, the efficiency increased by approximately 12% above the rated efficiency [23]. Mondol et al. [24] looked at a 13 kW PV array installed on a roof in

Northern Ireland and found the PV cell output to be approximately 10% less than rated during the summer season when the temperature was warm [25]. Another study reported on a 5.3 kW PV array installed on the East Coast of Saudi Arabia where the air temperature reached 60°C, and the resulting output was 35% less than rated efficiency [25, 26]. Thus, one would expect to see a significant increase in the solar PV *efficiency* for the solar PV arrays in Iqaluit where the ambient temperature is cold most of the year.

5.3 Solar PV Panel Efficiency

The solar PV panel efficiency is determined by dividing the electric DC power output by the input irradiance on the surface [27],

$$\eta = \frac{P}{AG_T}.$$

(5.1)

where P is the power output (W), A is the area of the array (m^2), and G_T is the solar irradiance on the tilted surface of the array (W/m^2). However, the efficiency of the solar PV panel is influenced by the PV cell temperature and irradiance, which can simply be estimated [27, 28],

$$\eta = \eta_{\text{ref}} \left[1 - \beta(T_C - T_{\text{ref}}) + \gamma \log \frac{G_T}{G_{\text{ref}}} \right],$$

(5.2)

where η_{ref} is the PV cell efficiency at standard reference conditions (G_{ref} is 1000 W/m^2 and T_{ref} is 298 K), γ and β are solar irradiance and temperature coefficients, respectively—these values are normally provided by the PV panel manufacturers. The solar irradiance coefficient (γ) is typically assumed to be zero; thus, Eq. 5.2 simplifies to Eq. 5.3 [27–29]:

$$\eta = \eta_{\text{ref}}[1 - \beta(T_C - T_{\text{ref}})]$$

(5.3)

5.4   Solar PV Cell Temperature (T_c)

Typically, for every 1°C rise above T_{ref}, the PV panel cell efficiency decreases by 0.25% for amorphous cells and 0.4–0.5% for crystalline cells [30]. These values can directly be used as the temperature coefficient β (%/K), in Eq. 5.3. PV manufacturers measure the electrical efficiency and temperature coefficient of the solar PV panel. Under Standard Reference Conditions (IEC 904–1 and IEC 60904–3), the solar PV panel is allowed to rest horizontally in the lab under electric lights, creating a simulated solar irradiance of 1000 W/m² on the PV cells with the ambient temperature set to a constant value (298 K for T_{ref}, others to determine β), and its current and voltage output are measured [31, 32].

An estimate of the maximum PV cell temperature is measured during the Nominal Operating Cell Temperature (NOCT) test 3 [27, 31]. This test is done on an open-rack PV panel under an open-circuit condition when the solar irradiance on the tilted surface is 800 W/m² with the PV panel tilted at 45° from the horizontal, at an ambient temperature of 293 K and air velocity of 1 m/s parallel to the panel. The solar PV cell temperature can be estimated from Eq. 5.4 [1, 27, 31, 33, 34],

$$T_c = T_a + (T_{NOCT} - T_{a,NOCT})\left(\frac{G_T}{G_{T,NOCT}}\right)\left(\frac{U_{L,NOCT}}{U_L}\right)\left(\frac{1-\eta}{\tau\alpha}\right), \qquad (5.4)$$

where T_a, T_{NOCT}, $T_{a,NOCT}$, $T_{T,NOCT}$, U_L, $T_{L,NOCT}$, τ and α are ambient temperature (K), nominal operating cell temperature (K), ambient temperature during the NOCT test (293 K), solar irradiance during the NOCT test (800 W/m²), overall heat loss coefficient (W/m²·K), overall heat loss coefficient during the NOCT test (W/m²·K), the transmittance of glazing, and absorptance of the PV cell, respectively.

5.4.1   Heat Losses from the Solar PV Arrays

After neglecting the conduction from the PV modules to the mounting structure, the overall heat loss coefficient can be estimated [27],

$$U_L = \frac{Q_{rad}}{A(T_C - T_a)} + h, \tag{5.5}$$

where Q_{rad} and h are the radiation heat loss (W) and the convection heat transfer coefficient (W/m$^2 \cdot$ K), respectively.

5.4.1.1 Radiation (Q_{rad})

According to Armstrong and Hurley, heat loss due to radiation from the PV arrays occurs from the PV top to the sky and the ambient air, and from the PV bottom to the ground, wall and ambient air, expressed as in Eq. 5.6 [1, 27, 35, 36].

$$Q_{rad} = Q_{rad,top} + Q_{rad,bottom} \tag{5.6}$$

It is also assumed that the temperatures of the covers on the top and bottom of the PV module are equal to the module's cell temperature and that the ground and wall temperatures are equal to the ambient temperature. Thus, radiation from both top and bottom can be estimated [1, 27, 35, 36, 37],

$$Q_{rad} = \varepsilon_1 \sigma F_1 A(T_C^4 - T_S^4) + \varepsilon_1 \sigma F_2 A(T_C^4 - T_{g/w}^4) + \varepsilon_2 \sigma F_3 A(T_C^4 - T_{g/w}^4), \tag{5.7}$$

where ε_1, $\varepsilon_2 \sigma$, F_1, F_2, F_3, T_S, and T_w are emissivity of the solar PV module at the top and bottom (0.91 and 0.85 at the top and bottom, respectively) [25, 33], Stefan–Boltzmann constant 5.67 × 10$^{-0.8}$ W/m$^2 \cdot$ K^4), view factors, sky temperature (K), ground temperature (K), and wall temperature (K), respectively.

5.4.1.1.1 The view factor (F)

The view factor (F) is the geometric fraction of the entire 180° that the solar PV array "sees," which is occupied by another body (sky, ground, or wall). Table 5.2 lists equations used to estimate the view factor from PV top to sky, PV top to ground and wall, and PV bottom to ground and wall, respectively [27, 36]. The view factor for PV bottom to ground and wall is the sum of Eqs. 5.10 and 5.11 in Table 5.1, resulting in a sum of 1 for façade-mounted arrays. Hence, $F_3 = 1$. In the case of a façade-mounted PV array, 90° of the 180° view is occupied by the sky, so the alternative view factors could be used: $F_1 = 0.5$, and $F_2 = 0.5$.

Table 5.1 View factor for radiative heat loss from a PV array installed at a given tilt angle [27, 36]

View factors location	Expression	
PV Top to Sky (F_1)	$\frac{1}{2}(1 + \cos\beta)$	(5.8)
PV Top to Ground and Wall (F_2)	$\frac{1}{2}(1 - \cos\beta)$	(5.9)
PV Bottom to Ground	$\frac{1}{2}[1 + \cos(180° - \beta)]$	(5.10)
PV Bottom to Wall	$\frac{1}{2}[1 - \cos(180° - \beta)]$	(5.11)

Note: β is the tilt angle of the array from horizontal.

5.4.1.1.2 Sky temperature (T_S)

There are numerous models available for estimating the sky temperature. The Swinbank model provides the sky temperature using only the local ambient temperature as an input. Thus, the sky temperature (K) can be estimated from Eq. 5.12 [1, 27, 35, 38].

$$T_S = 0.0552\, T_a^{1.5} \tag{5.12}$$

Bliss developed an equation where the sky temperature is related to the water vapor content of the ambient air [39],

$$T_S = T_a \left[0.8 + \frac{T_{dp} - 273}{250} \right]^{0.25},$$

where T_{dp} is the dew point temperature (K).

5.4.1.2 Convection

According to Hurley and Armstrong (2010), the convective heat loss from the PV array occurs at its top and bottom surfaces with exchange taking place with the ambient air and can be characterized as [27, 35, 36],

$$Q = Q_{conv,top} + Q_{con,bottom} \tag{5.14}$$

$$Q_{conv,i} = h_i A (T_C - T_a) \tag{5.15}$$

where $Q_{conv,top}$ and $Q_{con,bottom}$ are the convective heat losses from top and bottom of the PV arrays to the ambient air (W), respectively.

Three relationships between the convective heat loss coefficient, h, and the wind speed, V_w (m/s), are shown in Eqs. 5.16, 5.17, and 5.18 below. Test et al. [40] collected their data outdoors using natural wind, and the wind speed was measured 1 m above the array.

$$h = 2.56V_w + 8.55 \qquad (5.16)$$

Charlesworth & Sharples (1998) measured the wind speed windward of the array [41].

$$h = 3.3V_w + 6.5 \qquad (5.17)$$

Sturrock & Cole (1977) measured the wind speed leeward of the array, blowing parallel to the long dimension of the array [42].

$$h = 5.7V_w \qquad (5.18)$$

5.4.1.3   Energy balance

At steady-state, the energy entering and leaving the PV array achieves equilibrium; that is, the PV array receives energy from sunlight (G_T) and there are losses of energy in the form of heat through radiation and convection, and in the form of electricity. The error in the energy balance is expressed [43],

$$\%\text{Error} = 100\left[G_T A - \sum (Q_{rad} + Q_{conv} + P) \right]/(G_T A) \qquad (5.19)$$

5.5   Methodology

5.5.1   PV Output Data

PV output power, ambient temperature, and wind speed data were acquired for a 2.86 kW array installed at Qulliq Energy Corporation (QEC) and a 10.4 kW array at the Arctic Winter Games

Arena (AWGA) for clear and sunny days in Winter (January 1), Spring (May 26) and Summer (July 2), 2017. The data used for the analyses was for the period 11:50 to 13:20 h on January 1 for both arrays, while on May 26 the period at QEC was from 10:00 to 20:40 h, and at AWGA, it was from 4:35 to 20:10 h. On July 2 at QEC, it was from 10:20 to 20:45 h, and at AWGA, it was from 4:40 to 19:45 h. Although data are available throughout the day, there were significant power fluctuations under low solar altitude conditions, thus making it difficult to analyze. A threshold of 300 W was implemented, such that only the time when the power (P) was greater than 300 W was considered in the analysis. For comparison, during the winter solstice, the average day in Iqaluit lasts approximately 4½ h, while it is approximately 20 h at the summer solstice.

Table 5.2 PV array characteristics [47, 48]

Location	Type of modules	Array area [m²]	Number of modules in array	Azimuth [°]	Module NOCT [K]	Temperature coefficient of maximum power, β [%/K]	Nominal maximum rated power [W]	Panel reference efficiency, η_R [%]
QEC	Jinko Solar JKM260pp-60 Poly-crystalline	18.0	11	45.0	318±2	0.42	260	15.89
AWGA	Canadian Solar CS6P-260 Poly-crystalline	64.3	40	11.3	318±2	0.41	260	16.16

The PV output power (DC) data for both locations were obtained from Fronius IG plus 10 kW Inverters (Wels, Austria), which are remotely monitored by Green Sun Rising Inc. (solar designers and contractors in Windsor, Ontario, Canada; personal communication), and the ambient temperature and wind speed data was obtained from Environment Canada [44]. Both ambient temperature and wind speed data were recorded at the Iqaluit International Airport, which is located approximately 2.6 km from QEC and 4.2 km from AWGA [45]. The wind speed was measured at the height of 10 m from the ground surface [46]. The arrays at both locations are façade-mounted with a tilt angle of 60°C (Green Sun Rising Inc.; personal communication) and are located approximately 2 km apart from each other [46]. Table 5.2 summarizes the PV panels used in these arrays.

5.5.2 Weather Data

Figure 5.1 shows the ambient temperature for January 1, May 26, and July 2, 2017, at hourly intervals. From the plot, it is observed that the coldest of the three days was January 1 when the temperature fluctuated between –17 and –23°C, while on May 26 and July 2, the temperature fluctuated between –5 and 5°C, and 7 and 16°C, respectively (Enviro. & Natural Resources, 2018). However, during the analysis period, the ambient temperatures ranged from –19°C to –21°C on January 1, while on May 26 and July 2 the ranges were 3°C to 5°C and 14°C to 15°C, respectively.

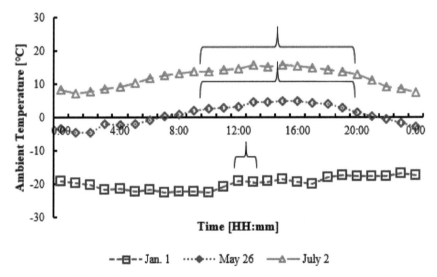

Figure 5.1 Ambient Temperature in Iqaluit for the days studied. The bracket shows the approximate analysis period at QEC.

Figure 5.2 shows the wind speed data for January 1, May 26, and July 2 at hourly intervals. The plot shows that on January 1, the wind speed fluctuated from 4 to 6 m/s, while for May 26 and July 2, it was fluctuating from 4 to 7 m/s for the analysis period (Enviro. & Natural Resources, 2018).

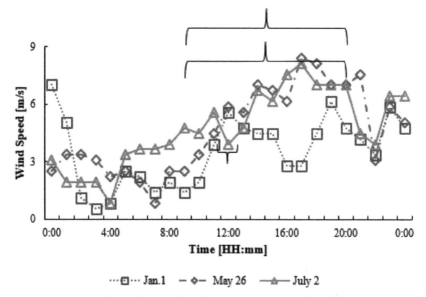

Figure 5.2 Wind speed in Iqaluit for the day studied. The bracket shows the approximate analysis period at QEC.

5.5.3   Estimation of PV Actual Output Efficiency

The PV array's output power data was obtained at 5 min intervals, while the ambient temperature and wind speed data were measured at 1 h intervals. Hence, the ambient temperature and wind speed data were interpolated to estimate values at 5 min intervals.

The PV arrays' estimated performance efficiencies and cell temperatures were calculated based upon the DC power and environmental weather using Eq. 5.1, and 3–5. Initially, the solar irradiance on the tilted array was estimated by using the reference efficiency in Eq. 5.1. Using the ambient temperature as the initial estimate of the PV cell temperature, the radiation heat loss from the array was estimated via Eq. 5.7, and convection heat loss by the array was estimated using Eq. 5.15. Then the cell temperature was estimated using Eq. 5.4. In the first iteration, the electrical efficiency was immediately recalculated by Eq. 5.3 using the new cell temperature. In subsequent iterations, the PV cell temperature was compared with the previous value, and the iterations stopped when it changed to less than 0.1 K. Figure 5.3 shows the calculation sequence. In addition, the following were assumed in performing the calculations:

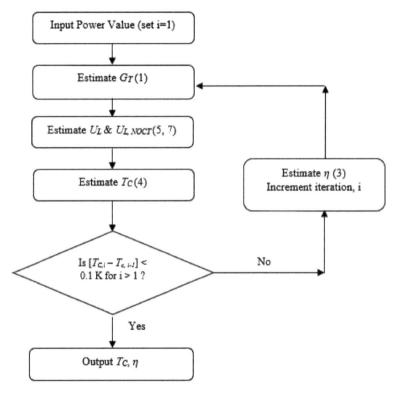

Figure 5.3 Flow chart showing the iterative calculation of T_C and η. Numbers in brackets are relevant equations.

1. Convection and radiation heat losses are taking place from the top and bottom of the PV arrays. The convection heat loss coefficient is the same for the top and bottom.
2. Convection and radiation heat losses from the edges and sides of the array are negligible.
3. $\tau\alpha = 0.9$ [34].
4. The glass temperature is equal to the PV cell temperature, and the ground and wall temperatures are equal to the ambient temperature [36].

5. Initially, the sky temperature was estimated using the Swinbank model (Eq. 5.12), the view factors from Table 5.1 were used, and Eq. 5.16 was used to calculate the heat transfer coefficient.

The overall heat loss coefficient $U_{L,NOCT}$ under the NOCT situation was estimated using the same equations as used to estimate the overall heat loss coefficient, U_L. However, a wind speed of 1 m/s was used at NOCT conditions, but when estimating the overall heat loss coefficient under field conditions (U_L) the measured wind speed was used. In order to separate the effect of the wind, another calculation was performed with a field wind speed of 1 m/s, effectively reducing the $U_L/(U_{L,NOCT})$ term in Eq. 5.4 to unity and eliminating the wind effect.

5.5.4 Sensitivity Analysis

A sensitivity analysis was performed to test the impact on the arrays' energy balances of the following assumptions/equations:

1. The convective heat transfer coefficient equations of Charlesworth & Sharples (Eq. 5.17) and Sturrock & Cole (Eq. 5.18) using the Swinbank sky temperature model (Eq. 5.12).
2. The three mentioned convective heat transfer coefficient equations (Eq. 5.16, 5.17, and 5.18) with the Bliss sky temperature model (Eq. 5.13).
3. The three mentioned convective heat transfer coefficient equations (Eq. 5.16, 5.17, and 5.18), and two sky temperature models (Eq. 5.12 and 5.13), with view factors $F_1 = F_2 = 0.5$.

5.6 Results and Discussion

Figures 5.4 and 5.5 show a plot of the arrays' estimated efficiencies on January 1, 2017 when both radiation to the sky and wind-induced convection to the ambient air are considered versus the case with radiation to the sky and wind-induced convection to the ambient with a speed of 1 m/s. A detailed analysis is shown for January 1 because it was the coldest day of the study period. When both convection to the ambient air and radiation to the sky were considered together, the estimated

PV efficiency was within the range of 18.7% to 18.8% at QEC, while at AWGA, it was 19.1%. This represents an 18% increase in output over that at the rated efficiency (~19% actual eff./~16% ref. eff.). However, when only radiation to the sky and convection at 1 m/s was considered, the estimated PV efficiencies dropped at both arrays by less than 0.5%, on average. Hence, the effect of convection to the ambient air was minimal on a winter day.

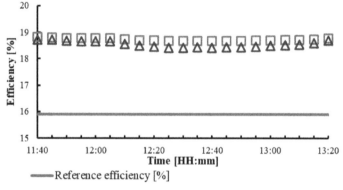

————Reference efficiency [%]

☐ Estimated Efficiency [%] [radiation to the sky and convection to ambient air]
△ Estimated Efficiency [%] [radiation to the sky and convection to ambient air when wind is 1 m/s]

Figure 5.4 Reference and estimated PV efficiencies. at QEC on January 1.

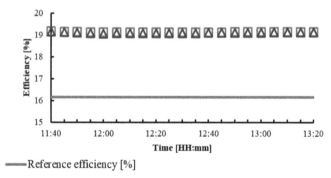

————Reference efficiency [%]

☐ Estimated Efficiency [%] [radiation to the sky and convection to the ambient air]
△ Estimated Efficiency [%] [radiation to the sky and convection to ambient air when wind is 1 m/s]

Figure 5.5 Reference and estimated PV efficiencies at AWGA on January 1.

Table 5.3 shows a summary of the estimated efficiencies and heat loss per area for the arrays at QEC and AWGA for the study period. On all three days under analysis, when both convection to the ambient air and radiation to the sky were considered, both arrays were generating above their reference efficiencies. At QEC, the estimated efficiency averaged 18.8% on the clear and sunny winter days, while on the clear and sunny spring and summer days, the efficiencies were 16.9% and 16.1%, respectively. At AWGA for the same days, the average efficiencies were 19.1%, 17.4%, and 16.7%, respectively. When only radiation to the sky and ambient air at 1 m/s was considered, it was found that at QEC, the estimated efficiency on the clear and sunny winter day averaged at 18.7%, while on the clear and sunny spring and summer days, the values were 16.0% and 15.2%, respectively. At AWGA for the same days, the average estimated efficiencies were 19.1%, 17.0%, and 16.4%, respectively.

Table 5.3 Estimated performance summary for both arrays

	QEC			AWGA		
Description	Jan. 1	May 26	July 2	Jan. 1	May 26	July 2
Mean Power [W]	584	1993	1851	758	3974	3384
Mean Temp. [°C]	−19.4	3.5	14.5	−19.4	3.5	14.5
Mean Wind speed [m/s]	4.9	6.3	6.3	4.9	6.3	6.3
PV rated η [%]	15.89	15.89	15.89	16.16	16.16	16.16
Mean Estimated η^* [%]	18.8	16.9	16.1	19.1	17.4	16.7
Mean Estimated η^{**} [%]	18.7	16.0	15.2	19.1	17.0	16.4
Mean Estimated η^{***} [%]	0.1	0.9	0.9	0	0.4	0.3
Mean q_{rad} [W/m^2]	73.6	107.4	109.1	47.3	88.3	84.0
Mean q_{conv} [W/m^2]	59.1	380.4	380.4	0	190.5	165.4

[*]mean estimated efficiency with both radiation to sky and convection to the ambient air.

[**]mean estimated efficiency with radiation to sky and convection to the ambient at 1 m/s.

[***]mean estimated efficiency difference between * and **.

$q_{rad.}$ is mean estimated radiation heat flow.

$q_{conv.}$ is mean estimated convection heat flow.

In addition, the estimated impact of convection to the ambient air on the estimated PV efficiencies was found to be below 1% in absolute terms on all three days for both arrays. The effect of the convection to ambient air on the PV efficiencies was found by subtraction. At QEC, the convection to the ambient air ranged from 0.1% on the clear and sunny winter day to 0.9% on the clear and sunny spring and summer days, while at AWGA, there was no impact on the clear and sunny winter day and 0.4% and 0.3% on the clear and sunny spring and summer days of the study period. Thus, the enhancement in efficiency at QEC resulted in an output that was 1.18 times the rated value on January 1. Similarly, AWGA experienced an 18% relative increase in output power for January 1.

5.6.1   Energy Balance

5.6.1.1   Base case

Figures 5.6 and 5.7 show a plot of the mean estimated input and output energy flows for the arrays at QEC and AWGA, respectively, for the analysis period. On all three days, the mean estimated energy output was less than the mean estimated energy input at both arrays, thus resulting in a positive error in the estimated energy balance (Eq. 5.19). At QEC, the error ranged from 8% to 9%, while at AWGA, it was from 3% to 6%. On the clear and sunny winter day, radiation to the sky was the more dominant heat loss mode, while on the clear and sunny spring and summer days, it was convection to the ambient air. The reason for the sudden flip from radiation to convection was primarily due to a significant rise in temperature difference between the PV cell and ambient air (ΔT) and a marginal increase of the convective heat transfer coefficient (h) due to the increase in induced wind speed from a mean of 4.9 m/s on the winter day to 6.3 m/s on the spring and summer days. Figure 5.8 shows graphically how the heat loss elements from QEC array changed during the seasons. The radiation temperature difference (RTD = $T_c^4 - T_s^4$) increased by less than twice from winter to spring and summer; however, the change in $\Delta T = -T_C - T_\alpha$ was greater than seven times. At AWGA, the situation is similar to

QEC. Some heat flows that were neglected in this analysis that may have caused the mean estimated input energy flow to be greater than the output are: conduction heat loss from the PV panels to the mounting brackets, and radiation and convection from the sides (edges) of the PV panels.

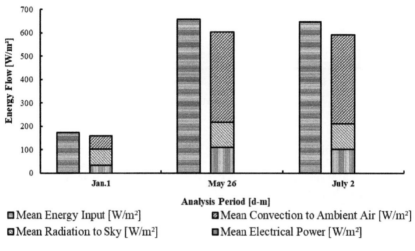

Figure 5.6 Energy balance at QEC during the analysis period

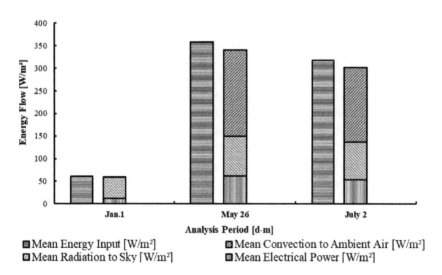

Figure 5.7 Energy balance at AWGA during the analysis period.

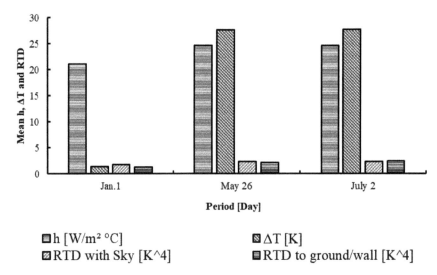

□ h [W/m² °C]   ⊠ ΔT [K]
▨ RTD with Sky [K^4]   ▤ RTD to ground/wall [K^4]

Figure 5.8 Mean change in h, ΔT, and RTD at QEC for analysis period. RTD value has been multiplied by 10^9.

5.6.1.2 Sensitivity

Figures 5.9, 5.10, 5.11, and 5.12 show plots of the estimated mean energy balance error (Eq. 5.19) for both QEC and AWGA arrays for the analysis period for the conditions outlined earlier under Section 5.3 (Sensitivity Analysis). In Fig. 5.9, Charlesworth & Sharples (Eq. 5.17) and Sturrock & Cole (Eq. 5.18) convective heat transfer coefficient models were tested against the base case convective heat transfer coefficient formula (Test et al.), using the Swinbank sky temperature model (Eq. 5.12). At QEC, Charlesworth & Sharples gives an estimated error ranging from 14% to 16%, while for Sturrock & Cole the error ranges from 32% to 36%. These are both more than the base case (Test et al.) which has an error of 8% to 9%. At AWGA the estimated error ranges from 3% to 12% and 3% to 32% for Charlesworth & Sharples and Sturrock & Cole, respectively, and again, the Test et al. model had the lowest error. The lower error indicates a more accurate convection heat transfer model, which implies a more accurate estimate of PV cell temperature and efficiency.

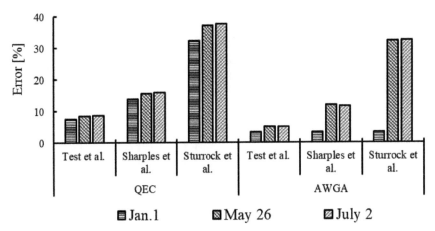

Figure 5.9 Energy balance error at QEC and AWGA for the analysis period (Swinbank sky temperature model and three convective heat transfer models).

In Fig. 5.10, the estimated energy balance errors for all three convective heat transfer coefficient models (Eq. 5.16, 5.17, and 5.18) were calculated using the Bliss model (Eq. 5.13) to estimate sky temperature. At QEC, the estimated error was 9% when using the Test et al. model, which is similar to the result using the Swinbank sky model. At AWGA for the same period and sequence, the estimated error ranged from –23% to 6% when using the Test et al. model. Negative errors indicate that the mean estimated energy input is less than the mean estimated energy output, which would not result from neglecting selected heat transfer losses.

In Fig. 5.11, the estimated energy balance errors for all three convective heat transfer coefficient models (Eq. 5.16, 5.17, and 5.18) were calculated using the Swinbank sky model (Eq. 5.12) and array view factors (F_1 and F_2) of 0.5. At QEC, using the Test et al. model, the estimated error was 8% to 9%. While at AWGA for the same period and sequence, the estimated errors ranged from 7% to 28%. Hence, when compared to the base case models, there was no reduction in the estimated error with the new view factors, and for AWGA the error was considerably higher.

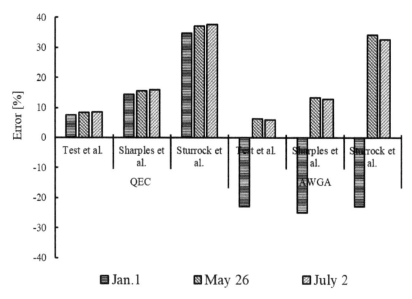

Figure 5.10 Energy balance error at QEC and AWGA for the analysis period (Bliss sky temperature model and three convective heat transfer models).

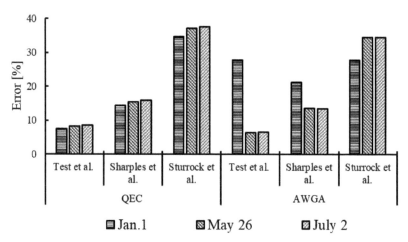

Figure 5.11 Energy balance error at QEC and AWGA for the analysis period ($F_{\text{top-sky}}$ = 0.5, Swinbank sky temperature model and three convective heat transfer models).

In Fig. 5.12, the estimated energy balance errors for all three convective heat transfer coefficient models (Eq. 5.16, 5.17, and 5.18) were calculated using the Bliss sky model (Eq. 5.12) and array view factors of 0.5. At QEC, using the Test et al. model, the estimated error was 9%, which is similar to the base case. While at AWGA, the estimated errors ranged from 7% to 28%. Hence, when compared to the base case models, it was found that the base case models had the least estimated error.

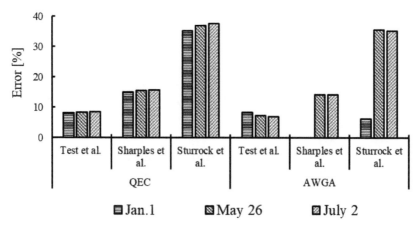

Figure 5.12 Energy balance error at QEC and AWGA for the analysis period ($F_{\text{top-sky}}$ = 0.5, Bliss sky temperature model and three convective heat transfer models).

Therefore, the base case models (Test et al. convective heat transfer coefficient, Swinbank sky temperature, and Armstrong and Hurley view factors) yield the least estimated energy balance error, thus are considered the most accurate in terms of estimating the energy balance for an array located in a cold climate.

5.6.2 Mean Annual Average

Tables 5.4 and 5.5 show a summary of the estimated performance and energy-weighted efficiency enhancements for both arrays for a clear and sunny day in each month of 2017. The results indicate that both arrays are performing above their rated efficiencies in this climate. The relative efficiency enhancement (REE) was calculated using Eq. 5.20.

$$REE = \left(\frac{\eta_{\text{daily mean est.}} - \eta_{\text{rated}}}{\eta_{\text{rated}}} \right) \times 100\% \qquad (5.20)$$

Arithmetically averaging the twelve values of *REE*, it was found that at QEC, the annual mean relative enhancement efficiency was 10%, while for the same period at AWGA, it was 11%. However, a more meaningful *energy-weighted average* (REE$_{\text{ewa}}$) would take into account the energy produced in each month. This was calculated using Eq. 5.21.

$$REE_{\text{ewa}} = \left(\frac{\sum_{i=1}^{12} E_i \times REE_i}{\sum_{i=1}^{12} E_i} \right) \times 100\%, \qquad (5.20)$$

where E_i is the mean energy (Wh) and REE (%) is the relative efficiency enhancement for the clear and sunny day in each month of 2017.

The mean annual energy-weighted efficiencies were 4% at QEC and 7% at AWGA.

Table 5.4 Calculation of monthly energy-weighted relative efficiency enhancement at QEC (reference η = 15.89%)

Description	Months											
	Jan. 1	Feb. 2	Mar. 6	Apr. 4	May 26	June 1	July 2	Aug. 15	Sept. 18	Oct. 1	Nov. 3	Dec. 5
$\eta_{\text{daily mean est.}}$ [%]	18.8	19.6	18.8	17.6	16.9	17.0	16.1	15.9	16.5	17.1	17.7	17.9
REE [%]	18.3	23.3	18.3	10.8	6.4	7.0	1.3	0.0	3.8	7.6	11.4	12.6
REE$_{\text{ewa}}$ [%]	1.0	1.0	9.0	13.0	8.0	8.0	1.0	0.0	3.0	6.0	3.0	0.0

Table 5.5 Calculation of monthly energy-weighted relative efficiency enhancement at AWGA (reference η = 16.16%)

Description	Months											
	Jan. 1	Feb. 2	Mar. 6	Apr. 4	May 26	June 1	July 2	Aug. 15	Sept. 18	Oct. 1	Nov. 3	Dec. 5
$\eta_{\text{daily mean est.}}$ [%]	19.1	19.8	19.3	18.3	17.4	17.6	16.7	16.5	17.1	17.4	18.2	18.1
REE [%]	18.2	22.5	19.4	13.2	7.7	8.9	3.3	2.1	5.8	7.7	12.6	12.0
REE$_{\text{ewa}}$ [%]	1.0	0.0	1.0	20.0	13.0	16.0	5.0	3.0	8.0	11.0	6.0	2.0

Figure 5.13 shows a plot of the estimated efficiencies for both arrays (QEC and AWGA) for a clear and sunny day in each month of the year (January to December 2017) versus literature efficiencies mentioned in Section 2. As the ambient temperature becomes colder, the estimated efficiency for both arrays increased compared to their reference efficiencies. However, the slope of the literature results is greater which reflects the fact that monocrystalline PV modules have a greater temperature co-efficient than polycrystalline modules, as exist at QEC and AWGA.

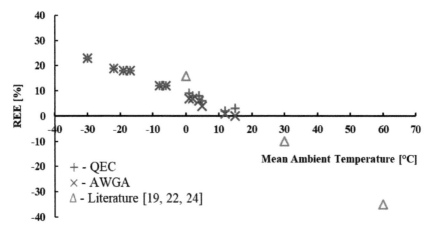

Figure 5.13 Mean monthly relative efficiency enhancements for both arrays and mean relative efficiency enhancements for literature projects.

5.7 Conclusion and Future Plans

These analyses estimated the output performance of the two PV arrays in Arctic conditions on sunny days in 2017. However, the enhancement in efficiency due to cold, windy Arctic conditions occurs more in the wintertime, when the available solar resource is less; thus, an energy-weighted average of the relative energy enhancement was calculated. Based on the estimated results, it can be concluded that both arrays are performing above their rated capacity by 4% to 7% on a mean annual energy-weighted basis. An energy balance was performed, considering radiation to the sky and convection to ambient air. During the winter day, radiation to the sky was the dominant heat loss mode in cooling of the PV cells, while during the other seasons, convection to the ambient air was the dominant heat loss mode.

In calculating the heat loss from the PV array, it was found that the Test et al. convection heat transfer coefficient model, the Swinbank sky temperature model, and the Armstrong and Hurley view factors provided the least error in the energy balance. The lower error gives more confidence that all significant heat losses were considered. Future work will estimate the horizontal solar irradiance (G) at a location based on measured power output from multiple PV arrays with different orientations.

References

1. S. Kalogirou (2009), Solar Energy Engineering, in *Processes and Systems*, Elsevier Inc., Burlington, MA.

2. L. Yang, X. Gao, F. Lv, X. Hui, L. Ma, X. Hou (2017), Study on the local climatic effects of large photovoltaic solar farms in desert areas, *Solar Energy*, 144, pp. 244–253.

3. M. Bayrakci, Y. Choi, J. R. S Brownson (2014), Temperature dependent power modeling of photovoltaics, *Energy Procedia*, 57, pp. 745–754.

4. M. Habiballahi, M. Ameri, S. H. Mansouri (2015), Efficiency improvement of photovoltaic water pumping systems by means of water flow beneath photovoltaic cells surface, *J. Solar Energy Eng.*, 137, p. 4044501.

5. T. Ma, W. Gu, L. Shen, L., M. Li (2019), An improved and comprehensive mathematical model for solar photovoltaic modules under real operating conditions, *Solar Energy*, 184, pp. 292–304.

6. K. Earley (2017), Why renewables are winning the 'carbon war', *Renew. Energy Focus*, 19–20, pp. 117–120.

7. B. Bora, R. Kumar, O. S. Sastry, B. Prasad, S. Mondal, A. K. Tripathi (2018), Energy rating estimation of PV module technologies for different climatic conditions, *Solar Energy*, 174, pp. 901–911.

8. S. Ghosh, V. K. Yadav, V. Mukherjee (2019), Impact of environmental factors on photovoltaic performance and their mitigation strategies–A holistic review, *Renew. Energy Focus*, 28, pp. 153–172.

9. J. M. Pearce, S. V. Obydenkova (2016), Technical viability of mobile solar photovoltaic systems for indigenous nomadic communities in northern latitudes, *Renew. Energy*, 89, pp. 253–267.

10. L. Dignard-Bailey, S. Martel, M. M. D. Ross (1998), Photovoltaics for the North: A Canadian Program, 2nd World Conference and Exhibition on Photovoltaic Solar Energy Conversion, Vienna, Austria.

11. M. Ross, Photovoltaics in cold climates, in: James & James, Science Publishers Ltd, 1999, pp. 16–20.

12. W. Tian, Y. Wang, J. Ren, L. Zhu (2007), Effect of urban climate on building integrated photovoltaics performance, *Energy Conversion Manag.*, 48, pp. 1–7.

13. N. Bowman, S. Shaari (2002), Photovoltaics in buildings: A case study for rural England and Malaysia, *Renew. Energy*, 15, pp. 558–561.

14. E. Skoplaki, A. G. Boudouvis, J. A. Palyvos (2008), A simple correlation for the operating temperature of photovoltaic modules of arbitrary mounting, *Solar Energy Mater. Solar Cells*, 92, pp. 1393–1402.

15. P. Trinuruk, C. Sorapipatana, D. Chenvidhya (2009), Estimating operating cell temperature of BIPV modules in Thailand, *Renewable Energy*, 34, pp. 2515–2523.

16. J. L. Balenzategui, M. C. Alonso García (2004), Estimation of photovoltaic module yearly temperature and performance based on nominal operation cell temperature calculations, *Renew. Energy*, 29, pp. 1997–2010.

17. E. Skoplaki, J. A. Palyvos (2009), On the temperature dependence of photovoltaic module electrical performance: A review of efficiency/power correlations, *Solar Energy*, 83, pp. 614–624.

18. E. Skoplaki, J. A. Palyvos (2009), Operating temperature of photovoltaic modules: A survey of pertinent correlations, *Renew. Energy*, 34, pp. 23–29.

19. Latitude, Iqaluit Nunavut Canada Latitude. http://latitude.to/articles-by-country/ca/canada/1534/iqaluit, (accessed 7 May, 2018).

20. M. Mussard (2017), Solar energy under cold climatic conditions: A review, *Renew. Sustain. Energy Rev.*, 74, pp. 733–745.

21. L. S. Pantic, T. M. Pavlović, D. D. Milosavljević, I. S. Radonjic, M. K. Radovic, G. Sazhko (2016), The assessment of different models to predict solar module temperature, output power and efficiency for Nis, Serbia, *Energy*, 109, pp. 38–48.

22. A. Kasaeian, Y. Khanjari, S. Golzari, O. Mahian, S. Wongwises (2017), Effects of forced convection on the performance of a photovoltaic thermal system: An experimental study, *Exp. Thermal Fluid Sci.*, 85, pp. 3–21.

23. Z. Peng, M. R. Herfatmanesh, Y. Liu (2017), Cooled solar PV panels for output energy efficiency optimisation, *Energy Conversion Manag.*, 150, pp. 949–955.

24. J. D. Mondol, Y. Yohanis, M. Smyth, B. Norton (2006), Long term performance analysis of a grid connected photovoltaic system in Northern Ireland, *Energy Conversion Manag.*, 47(18), pp. 2925–2947.

25. A. Vasel, F. Iakovidis (2017), The effect of wind direction on the performance of solar PV plants, *Energy Conversion Manag.*, 153, pp. 455–461.

26. I. El-Amin, S. Rehman (2012), Performance evaluation of an off-grid photovoltaic system in Saudi Arabia, *Energy*, 46(1), pp. 451–458.

27. F. Fouladi, P. Henshaw, P. D. S-K. Ting (2013), Enhancing smart grid realisation with accurate prediction of photovoltaic performance based on weather forecast, *Int. J. Environ. Stud.*, 70(5), pp. 754–764.

28. Spectrolab Inc. Sylmar C. A (1977), Photovoltaic Systems Concept Study: Final Report (ALO-2748-12), Springfield, VA: US Department of Energy, Division of Solar Energy.

29. D. L. Evans (1981), Simplified method for predicting photovoltaic array output, *Solar Energy*, 27(6), pp. 555–560.

30. H. Sainthiya, N. S. Beniwal, N. Garg (2018), Efficiency improvement of a photovoltaic module using front surface cooling method in summer and winter conditions, *J. Solar Energy Eng.*, 140(6), p. 061009.

31. International Electrotechnical Commission (1993), *Crystalline Silicon Terrestrial Photovoltaic (PV) Modules – Design Qualification and Type Approval*, 2nd ed, Geneva: IEC, International Standard EN-61215, 1993-04.

32. B. Hüttl, L. Gottschalk, S. Schneider, D. Pflaum, A. Schulze (2019), Accurate performance rating of photovoltaic modules under outdoor test conditions, *Solar Energy*, 177, pp. 737–745.

33. E. Rossi, H. Ossenbrink (1992), European solar test installation: Qualification test procedures for crystalline silicon photovoltaic modules (ISSN 1018-5593), Commission of the European Communities.

34. N. Aste, G. Chiesa, F. Verri, Design (2008), Development and performance monitoring of a photovoltaic-thermal (PVT) air collector, *Renew. Energy*, 33-5, pp. 915–927.

35. J. A. Duffie, W. A. Beckman, W. A. (1980), *Solar Engineering of Thermal Processes*, 2nd ed., John Wiley and Sons, New York.

36. S. Armstrong, W. G. Hurley (2010), A thermal model for photovoltaic panels under varying atmospheric conditions, *Appl. Thermal Eng.*, 30(11–12), pp. 1488–1495.

37. G. Notton, C. Cristofari, M. Mattei, P. Poggi (2005), Modelling of a double-glass photovoltaic module using finite differences, *Appl. Thermal Eng.*, 25, pp. 2854–2877.

38. W. C. Swinbank. (1963), Long-wave radiation from clear skies, Quarterly *J. R. Meteorol. Soc.*, 89, pp. 339–348.

39. R. W. Bliss (1961), Atmospheric radiation near the surface of the ground: A summary for engineers, *Solar Energy*, 5(3), pp. 103–120.

40. F. L. Test, R. C. Lessmann, A. Johary (2009), Heat transfer during wind flow over rectangular bodies in the natural environment, *J. Heat Transfer*, 103, pp. 262–267.

41. P. S. Charlesworth, S. Sharples (1998), Full-scale measurements of wind-induced: Convective heat transfer from a roof mounted flat plate solar collector, *Solar Energy*, 62(2), pp. 69–77.

42. N. S. Sturrock, R. J. Cole (1977), The convective heat exchange at the external surface of buildings, *Building Environ.*, 12, pp. 207–214.

43. M. K. Fuentes (1987), A simplified thermal model for flat-plate photovoltaic arrays, Sandia Report (SAND85-0330) UC-63.

44. Environment and Natural Resources, Gov't of Canada, Hourly data report for Iqaluit Climate, 2017. http://climate.weather.gc.ca/climate_data/hourly_data_e.html?hlyRange=2004-12-16%7C2018-06-11&dlyRange=2004-05-25%7C2018-06-10&mlyRange=2005-03-01%7C2007-11-01&StationID=42503&Prov=NU&urlExtension=_e.html&searchType=stnName&optLimit=yearRange&StartYear=1840&EndYear=2013&selRowPerPage=25&Line=2&searchMethod=contains&Month=1&Day=31&txtStationName=iqaluit&timeframe=1&Year=2017 (accessed 4 June, 2018).

45. Google Maps, Map of Iqaluit, Nunavut, Canada showing direction. https://www.google.com/maps/dir///@42.2926058,-83.0770962,15z (accessed 18 July, 2018).

46. J.-P. Pinard, Potential for Wind Energy in Nunavut Communities. https://www.qec.nu.ca/sites/default/files/potential_for_wind_

energy_in_nunavut_communities_2016_report_0.pdf, 2016 (accessed 8 July, 2018).

47. Jinko Solar, JKM260PP-60 Poly crystalline module 240-260 Watts. https://www.jinkosolar.com/ftp/EN-Eagles-260PP_v1.0_rev2013.pdf, 2013 (accessed 7 May, 2018).

48. C.S.I., ClearPower CS6P-260/265P-SD. https://www.canadiansolar.com/downloads/datasheets/v5.4/Canadian_Solar-Datasheet-CS6PPSD_SmartDC-v5.4en.pdf, 2016 (accessed 7 May, 2018).

Chapter 6

A Review of Current Development in Photovoltaic-Thermoelectric Hybrid Power Systems

Xi Wang, Paul Henshaw, and David S.-K. Ting

Turbulence and Energy Laboratory, University of Windsor, Windsor, Ontario, Canada

henshaw@uwindsor.ca

The photovoltaic-thermoelectric (PV-TE) hybrid system is a promising clean energy technology. A PV cell cannot convert all photons into electrical power. Absorbed unconverted photons increase the PV cell temperature, weakening its energy conversion efficiency. Through the thermoelectric effect, a thermoelectric generator (TEG) can convert part of the thermal energy into electricity, improving the overall performance of the PV-TE hybrid system. This chapter introduces the working principle and categories of PV-TE hybrid systems. It comprehensively reviews the development, structural features, and optimization of PV-TE systems and aims to serve as a useful reference of PV-TE hybrid technology.

Climate Change and Pragmatic Engineering Mitigation
Edited by Jacqueline A. Stagner and David S.-K. Ting
Copyright © 2022 Jenny Stanford Publishing Pte. Ltd.
ISBN 978-981-4877-97-8 (Hardcover), 978-1-003-25658-8 (eBook)
www.jennystanford.com

Nomenclature

A_N	cross-sectional area of the N-type semiconductor, mm^2
A_P	cross-sectional area of the P-type semiconductor, mm^2
$DSSC$	dye-sensitized solar cell
IAI	ITO/Ag/ITO
I_{PV}	current produced by a PV cell, A
IR	infrared radiation
I_{TEG}	current produced by a TEG, A
ITO	indium tin oxide
K	net thermal conductance, W$\cdot$K^{-1}
k_N	thermal conductivity of the N-type semiconductor, W$\cdot$m$^{-1}\cdot$K^{-1}
k_p	thermal conductivity of the P-type semiconductor, W$\cdot$m$^{-1}\cdot$K^{-1}
L_N	length of the N-type semiconductor, mm
L_p	length of the P-type semiconductor, mm
$MCHP$	micro-channel heat pipe
$NCPV$	nanofluid-cooled photovoltaic
NIR	near infrared ray
PV	photovoltaic
$PV\text{-}TE/T$	photovoltaic-thermoelectric/thermal
P_{PV}	output power produced by a PV cell, W
P_{TEG}	output power produced by a TEG, W
$\dot{Q}_F$	Fourier heat transfer rate, W
$\dot{Q}_{input}$	input heat transfer rate, W
$\dot{Q}_J$	Joule heat transfer rate, W
$\dot{Q}_P$	Peltier heat transfer rate, W
R_i	internal resistance, Ω
R_L	load resistance, Ω
S	Seebeck coefficient, V$\cdot$K^{-1}
SSA	solar selective absorber
TE	thermoelectric
TEG	thermoelectric generator
T_c	temperature of the cold side, K
T_h	temperature of the hot side, K
UV	ultraviolet radiation

V_{PV}	voltage produced by a PV cell, V
V_{TEG}	voltage produced by a TEG, V
ΔT	temperature different, K
η_{PV}	efficiency of a PV cell
η_{TEG}	efficiency of a TEG
ρ_N	electrical resistivity of the N-type semiconductor, $\Omega \cdot mm$
ρ_P	electrical resistivity of the P-type semiconductor, $\Omega \cdot mm$

6.1   Introduction

Modern civilization has achieved exceptionally rapid development over the past decades. Meanwhile, the modern lifestyle drives an increasing number of demands for energy. The global energy demand in the 1980s was about 10 TW and increased five-fold by the early twenty-first century [1]. Nowadays, energy demand is experiencing an increase at a rate of about 2% per year [1]. It is estimated that the global energy demand in 2057 will be 3 times as much as the early twenty-first century [1]. There is no doubt that fossil fuel also plays a dominant role in the energy supply system [1]. However, relying on fossil fuel excessively will inevitably put humans in a dire state, due to its finite supply and environmental issues. Currently, exploiting renewable energy is considered a useful way to avoid the dilemma of supplying energy while minimizing environmental effects.

An increasing number of companies are willing to invest in renewable energy, such as wind, hydro and solar, making it possible for related technologies to develop at an astounding pace in recent years. Solar power is considered the renewable energy source with the most potential, as the solar radiation received by the earth's surface is about 1.2×10^5 TW which is four orders of magnitude greater than the global energy demand [2]. Besides, its excellent cost-effectiveness is the reason that solar power systems, especially photovoltaic (PV) systems, have been applied recently all over the world. By the end of 2019, the global installed capacity of PV systems reached 627 GW [3]. The International Energy Agency (IEA) predicts that the global installed capacity of PV systems will increase to 4.7 TW by 2050 [4].

Although there is an expected growth in the number of PV systems, the temperature problem is one of the unavoidable challenges, which may impede the further development of PV technology. The wide incident solar spectrum ranges from 280 to 4000 nm [5]. However, it is impossible for PV cells to convert the whole spectrum of radiant energy to electrical power. At present, the silicon thin film solar cell is considered as one of the promising PV technologies with a theoretical efficiency limit of about 30% [5]. Even at this efficiency, there is a significant fraction of the incoming solar energy that is converted into heat, which can lead to a temperature rise in PV cells. Normally, a temperature rise of 1°C can lead to a decrease of 0.08–0.45% in the efficiency (or output power) of PV cells [6–8]. Traditionally, cooling systems have not been used for PV systems. However, various thermal enhancement devices, involving active and passive cooling methods, have been developed to reduce the PV cell temperature [9].

Compared with traditional cooling schemes, however, it is more attractive to add a device to the PV system to partially recover the waste heat. The energy recovery device can not only decrease the temperature of the PC cell surface but also improve the overall efficiency for the power system. Among the numerous energy recovery technologies, the thermoelectric generator (TEG) is considered as an excellent option for PV systems. Similar to PV panels, TEGs are a kind of semiconductor energy convertor, but driven by a temperature difference to produce electric power [10]. As a kind of solid energy device, TEGs are equipped with many outstanding features. Firstly, the TEG has a compact structure [11–12]. In this way, it is easy for PV cells and TEGs to make a hybrid power system through a simple combination. Besides, there are not any moving components nor chemical reactions, making the TEG work safely and reliably [11–12]. One important thing is that the usage of the TEG device poses no threat to the surroundings as it has zero emission while operating [5]. With the continuous development of the TE material, the efficiency of TEGs have improved (the efficiency of a commercial TEG module can reach about 10%) [13]. Therefore, hybrid power systems, consisting of PV cells and TEGs is worthy of research.

A review of the current status of PV-TE hybrid systems is reported in this paper. The review starts from the illustration of the PV-TE hybrid system working principle. Then, some representative PV-TE systems reported in recent years are introduced, including their structure, working principle and characteristics. Finally, an overview of the main challenges for PV-TE system development and methods to optimize the performance of the hybrid system are presented. Overall, this chapter provides readers with a useful summary about the development and prospects for PV-TE hybrid systems.

6.2 Working Principles of PV-TE System

6.2.1 Photovoltaic Cells

A PV cell is a kind of solid energy device, which can utilize the photoelectric effect to convert sunlight into electrical power directly without any moving components or chemical reactions. In this way, it is possible for PV cells to operate reliably for a long-term, and without any exhausted gas emission [14].

Normally, a PV cell is made up of two thin layers of semiconductors. These thin layers form a P-N junction. Due to the characteristics of semiconductors, it is easy for the valence electrons to be excited by photos from sunlight. If the photon energy is greater than the band gap of the semiconduction material, these excited electrons will jump to the conduction band, forming free electrons [1]. In this way, under an internal electric field across the PV material, the free electrons will flow through the external circuit. Figure 6.1 displays the schematic diagram of the photovoltaic cell.

The output power, $P_{PV}(W)$, can be calculated by Eq. (6.1) [1, 15].

$$P_{PV} = V_{PV}I_{PV},\tag{6.1}$$

in which I_{PV} and V_{PV} are current (A) and voltage (V) produced by a PV cell.

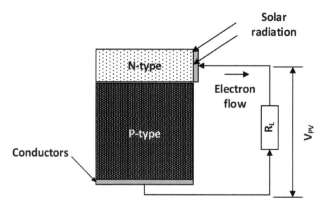

Figure 6.1 Schematic diagram of a PV cell.

The efficiency of the PV cell, η_{PV}, is a ratio between output power and energy input, as shown in Eq. (6.2):

$$\eta_{PV} = \frac{P_{PV}}{\dot{Q}_{in}} \tag{6.2}$$

It is impossible for all wavelengths of solar energy to be converted into electrical power by the PV cell, which limits its conversion efficiency. Table 6.1 displays some PV materials' efficiency [1]. Figure 6.2 shows a schematic diagram of the solar spectrum used by PV cells. According to the figure, the cells only partially utilize the energy of photons greater than the band gap energy [1, 5]. Meanwhile, the electrical conversion efficiency of photons reaches to the maximum when the photons' energy is close to that of PV cell band gap [5, 16]. The energy in photons which is not converted to electricity will be dissipated as waste heat [16, 17]. Therefore, it is worthwhile for a PV system to add a device to recover the waste heat.

Table 6.1 Sample PV materials' efficiency

PV material	Conversion efficiency	Ref.
Amorphous Si	6–7%	[18]
Cadmium telluride thin film	10–11%	[1]
Copper indium gallium selenide	10–13%	[1]
Polycrystalline Si	13–17%	[19]
Monocrystalline Si	14–20%	[18]

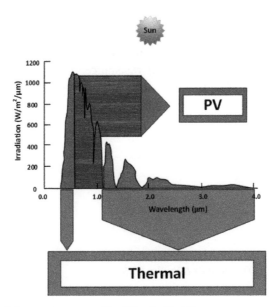

Figure 6.2 Solar spectrum used by PV cells [5, 16].

6.2.2 Thermoelectric Generator

Valence electrons in semiconducting materials can also be excited by thermal energy, producing thermoelectric effects, such as the Seebeck effect, Peltier effect, and Thomson effect [20]. The basic principle of the Seebeck effect is that charge carriers (holes in P-type semiconductor and electrons in N-type semiconductor) have a high population near the cold end of a semiconducting element. Hence, near the cold conductor, there is an excess population of holes in the P-type semiconductor and an excess population of electrons in the N-type semiconductor. This creates a potential for current flow. Normally, a TEG consists of three different parts (as shown in Fig. 6.3): TE couples, substrate, and conductors [20].

Based on the Seebeck effect, the open circuit voltage of the TEG, V_{TEG} (*V*), is a product of the Seebeck coefficient and temperature difference as shown in Eq. (6.3) [20, 21].

$$V_{TBG} = S\Delta T = S\,(T_h - T_c),\tag{6.3}$$

where, S is the Seebeck coefficient ($V \cdot K^{-1}$) mainly decided by the TE material; ΔT is the temperature difference (K) between the hot side temperature, $T_h(\text{K})$, and the cold side temperature, $T_c(\text{K})$.

Based on Ohm's law, the current, $I_{\text{TBG}}(A)$, is the ratio of the open circuit voltage and total resistance, as shown in Eq. (6.4) [20, 22, 23]. Meanwhile, the output power, $P_{\text{TBG}}(W)$, can be calculated by Eq. (6.5) [20, 22, 23].

$$I_{\text{TEG}} = \frac{V_{\text{TEG}}}{(R_i + R_L)} = \frac{S(T_h - T_c)}{(R_i + R_L)} \tag{6.4}$$

$$P_{\text{TEG}} = I_{\text{TEG}}{}^2 R_L = \left[\frac{S(T_h - T_c)}{(R_i + R_L)} \right]^2 R_L, \tag{6.5}$$

where R_L is the load resistance (Ω), and R_i is the internal resistance (Ω) calculated by Eq. (6.6).

$$R_i = \frac{\rho_P L_P}{A_P} + \frac{\rho_N L_N}{A_N} \tag{6.6}$$

in which A_P, L_P, and ρ_P are the cross-sectional area (mm^2), length (mm), and electrical resistivity ($\Omega \cdot$mm) of the P-type semiconductor; A_N, L_N, and ρ_N are the cross-sectional area (mm^2), length (mm), and electrical resistivity ($\Omega \cdot$mm) of the N-type semiconductor.

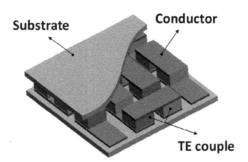

Figure 6.3 Structural diagram of the TEG module.

As was the case for a PV module, the efficiency of a TEG module is also the ratio of output power and energy input rate. Through analyzing the thermal process of the hot side boundary (as shown in Fig. 6.4), there are four ways that thermal energy passes through the boundary: energy input, Joule heat, Fourier heat, and Peltier heat [24].

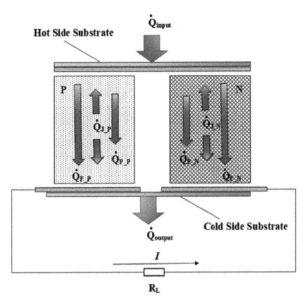

Figure 6.4 Schematic diagram of a TEG heat transfer process in hot side thermal boundary.

Energy input and Joule heat are absorbed by the boundary, and Fourier heat and Peltier heat are released from the boundary [24]. Specially, the Joule heat can be considered as an internal heat source, releasing the energy in all directions as a result of electron flow. In order to simplify the physical process, it is assumed that only half of the Joule heat is absorbed by the hot side thermal boundary [24, 25]. Therefore, according to the law of energy conservation, the energy input rate, $\dot{Q}_{in}$(W), is the algebraic sum of these heat loss rates from the hot side substrate which can be shown in Eq. (6.7) [24, 25].

$$\dot{Q}_{in} = \dot{Q}_F + \dot{Q}_P - \frac{1}{2}\dot{Q}_J, \tag{6.7}$$

where $\dot{Q}_F$, $\dot{Q}_P$, and $\dot{Q}_J$ are Fourier heat rate (W), Peltier heat rate (W), and Joule heat rate (W), which can be calculated by the following equations (Eqs. (6.8–6.10)) [25].

$$\dot{Q}_F = K(T_h - T_c) \tag{6.8}$$

in which K is the net thermal conductance $(W \cdot K^{-1})$ of the TE couple $\left(K = \dfrac{k_p A_p}{L_p} + \dfrac{k_N A_N}{L_N} \right)$; k_p and k_n are the thermal conductivities $(W \cdot mm^{-1} \cdot K^{-1})$ of the P and N-type semiconductors.

$$\dot{Q}_P = S T_h I_{TEG} = \frac{S^2 T_h (T_h - T_c)}{R_i + R_L} \tag{6.9}$$

$$\dot{Q}_J = I_{TEG}{}^2 R_i = \left[\frac{S(T_h - T_c)}{R_i + R_L} \right]^2 R_i \tag{6.10}$$

In this way, the efficiency of a TEG module, η_{TEG}, can be calculated through Eq. (6.11).

$$\eta_{TEG} = \frac{\left[\dfrac{S(T_h - T_c)}{R_i + R_L} \right]^2 R_L}{K(T_h - T_c) + \dfrac{S^2 T_h (T_h - T_c)}{R_i + R_L} + \dfrac{1}{2} \left[\dfrac{S(T_h - T_c)}{R_i + R_L} \right]^2 R_i} \tag{6.11}$$

Early TE materials led to an efficiency below 4%, impeding the wide application of the TEG technology for a long time [26]. However, with the development of better materials, TEG performance acquired an obvious improvement. The conversion efficiencies of the TE materials shown in Fig. 6.5 are more than 4%. And even, when the temperature difference reaches about 600 K, the efficiency of a Skutt/Bi$_2$Te$_3$ TEG is more than 10% [13].

Specially, some researchers have verified that compared with the poor energy conversion efficiency of a TEG, it normally has a better exergy efficiency (30%–40%) [27]. The exergy efficiency is used to evaluate the energy quality of a system. This means that TEG can be applied in some energy systems to decrease the irreversible heat loss.

PV-TE hybrid systems can increase the overall performance of the system.

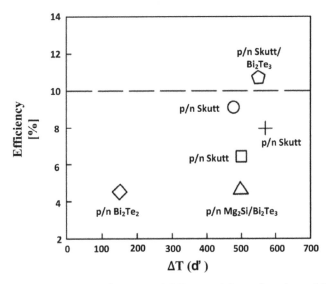

Figure 6.5 Conversion efficiency of TE materials under the cold side temperature fixed at 323.15 K [13].

6.2.3 Summary of PV-TE System Development within the Past Decade

The idea of PV-TE hybrid system was first introduced in 2008 by Tritt, who mainly focused on studies related the application of ultraviolet (UV) and infrared (IR) radiation [28, 29]. During the period of the more than a decade, there are a large number of PV-TE hybrid configurations proposed. Table 6.2 displays the main contributions for PV-TE hybrid systems in this decade.

Table 6.2 Summary of PV-TE system designs within the decade

Author	Technique	Contribution
2010: Leonov [30]	Integrated PV-TE system	Applied the PV-TE system in wearable devices
2012: Mizoshiri et al. [31]	Spectrum splitting PV-TE system; Thin-Film TE module	The open circuit voltage of the hybrid system was increased 1.3% compared to that of a pure PV system

(Continued)

Table 6.2 (*Continued*)

Author	Technique	Contribution
2015: Elsarrag et al. [28]	Spectrum splitting PV-TE system;	Through splitting the IR ray from the sunbeam, the power of the PV-TE system was increased around 10% compared with a PV system
2016: Da et al. [32]	Integrated PV-TE system	Utilized bionics techniques (a moth-eye nanostructure) to improve the efficiency of the PV-TE system more than 2%
2017: Sibin et al. [33]	Spectrum splitting PV-TE system;	Utilized ITO/Ag/ITO (IAI) multilayer coatings to improve transmittance of visible light and reflectance of IR
2017: Mohsenzadeh et al. [34]	Integrated PV-TE system	Designed a novel concentrating PV-TE system; its daily electrical power generation reached 43.36 W/m^2
2018: Marandi et al. [35]	Integrated PV-TE system	Utilized PV-TE hybrid modules to form a solar cavity, making its efficiency reach a peak value of 21.9%
2019: Li et al. [36]	Integrated PV-TE system; Micro-channel heat pipe (MCHP) array	Through the MCHP array, the efficiency of the PV-TE system can stay above 14%. Meanwhile, the temperature of the PV cell is no higher than above 20°C than the ambient temperature
2020: Zhou et al. [37]	Integrated PV-TE system	Utilized a thermal collector to increase the output power by 11.2% for a PV-TE system

Based on Table 6.2, the PV-TE hybrid designs can be divided into two different types, which are integrated systems and spectrum splitting systems.

6.3 Integrated PV-TE Hybrid System

The integrated PV-TE hybrid system is a method to efficiently utilize solar energy with a wide spectrum. Figure 6.6 is a schematic diagram of an integrated PV-TE hybrid system. A simple integrated PV-TE system mainly consists of three different parts, which are the PV panel, TEG module and cooling device [5]. Normally, the TEG module is placed between the PV panel and cooling device. The temperature difference produced from the PV panel and cooling device can drive the TEG module to partially recover the wasted thermal energy. As for the integrated PV-TE system, the heat sink consisting of the TEG module and cooling device can limit the temperature rise for the PV panel, making the PV cell work at an optimum temperature as soon as possible. Meanwhile, compared to the traditional PV system, the TEG module can convert the wasted heat produced from the PV cell partially into electrical power, which can improve the energy system performance further.

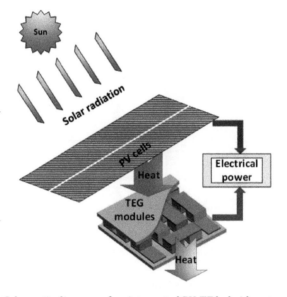

Figure 6.6 Schematic diagram of an integrated PV-TE hybrid system.

In the recent decade, many PV-TE systems were proposed. Marandi et al. [35] designed a solar cavity that consisted of PV-TE modules as shown in Fig. 6.7(a). Thereinto, there are five PV panels comprising the surface of the cavity. As for each PV panel, a cascade array with four TEG modules is combined with the back surface of the PV panel through a thermally conductive adhesive (as shown in Fig. 6.7(b)). The experimental test indicated that the efficiency of this hybrid system reached a peak value of about 21.2% [5, 35]. Besides, the experimental results indicated that, compared to a flat-PV-TE system (traditional system), the efficiency of this solar cavity was improved by 18.9% [35].

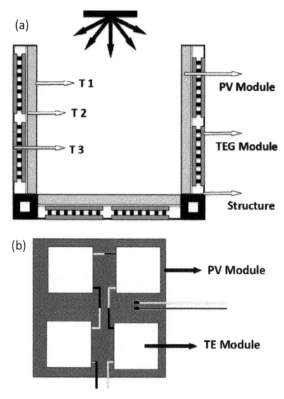

Figure 6.7 Schematic diagram of the PV-TE hybrid system, (a) structure of the solar cavity, and (b) structure of the PV-TE module [35].

Additionally, some advanced solar energy receivers were added into integrated PV-TE systems in order to improve their performance further. Da et al. [32] utilized a moth-eye nanostructure to design a novel integrated PV-TE system as shown in Fig. 6.8. This system is mainly made up of a concentrator, PV cell, TEG module, and cooling system. The PV cell is equipped with a GaAs nanostructure consisting of a moth-eye structure made by p-$Al_{0.8}Ga_{0.2}As$, heavily doped p-GaAs, a lightly doped n-GaAs, n-$Al_{0.3}Ga_{0.7}As$, and SiO_2 [32]. The heavily doped p-GaAs acts as the emitter, and the lightly doped n-GaAs acts as the base. The functions of the n-$Al_{0.3}Ga_{0.7}As$ and SiO_2 are back surface field and transmission enhancement. The moth-eye structured design can decrease the reflection of full-spectrum photons from sunbeams effectively. Meanwhile, this structure can enhance the transmission for those photons whose energy is lower than that of band gap of the PV cell, making it possible for the hybrid system to utilize the solar radiation more reasonably. Through the simulation, the efficiency of the hybrid system was improved over that of a PV alone system by more than 2% [32].

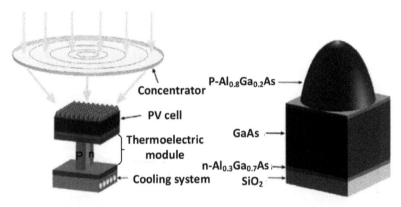

Figure 6.8 Schematic diagram of the PV-TE hybrid system with a moth-eye nanostructure [32].

In order to utilize the solar radiation efficiently, Zhao et al. [38] developed a broad-spectrum PV-TE system. Figure 6.9 shows that unlike most traditional PV-TE systems, a dye-sensitized solar cell (DSSC) and solar selective absorber (SSA) replaced the PV cell. The DSSC can convert the photons whose wavelengths are below 920 nm into electrical power. The SSA absorbs the sunlight with higher wavelengths (>920 nm) for conversion into thermal energy [38]. Meanwhile, the thermal energy released from the SSA can be used as a thermal source for TEG modules.

It is acknowledged that the infrared photons are associated with a strong thermal effect, leading to a temperature increase of a body surface without any possibility of supplying energy above the band gap. Based on the PV theory, the high temperature has negative effects on the performance of a PV cell. Because of this, some researchers tried to add a spectrum filter in the integrated PV-TE system. Zhou et al. [37] designed a photovoltaic-thermoelectric/thermal (PV-TE/T) system as shown in Fig. 6.10. The thermal collector is made up of a spectrally selective absorbing fluid, which can reduce the infrared content of sunlight.

During system operation, the PV cell can utilize the filtered sunlight to generate power; meanwhile, the TEG module can partially recover the wasted heat from the PV cell.

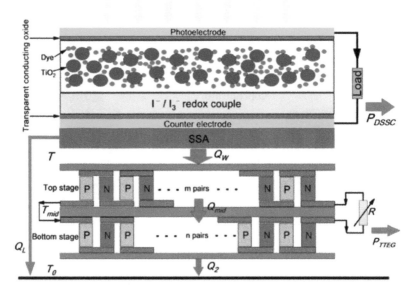

Figure 6.9 Schematic diagram of the broad-spectrum PV-TE hybrid system [38].

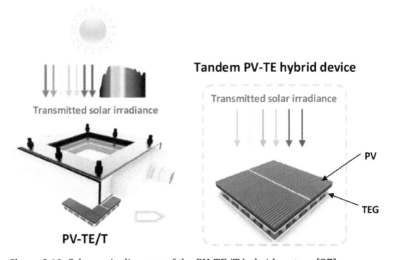

Figure 6.10 Schematic diagram of the PV-TE/T hybrid system [37].

Some researchers paid more attention to the heat exchange system enhancement for integrated PV-TE hybrid systems. Two representative studies were done by Lekbir et al. [39], and Li et al. [36]. They applied a nanofluid-based cooling device or a micro-channel heat pipe into the hybrid system, respectively. Hence, the two systems were named NCPV-TE and PV-TE-MCHP systems, as shown in Figs. 6.11(a) and 6.11(b). The basic idea is to utilize advanced heat exchange technology to transfer the wasted heat released from the PV cell to the TEG module, which can increase the temperature difference on both sides of the TEG, and thus enhance the heat recovery rate for the hybrid system.

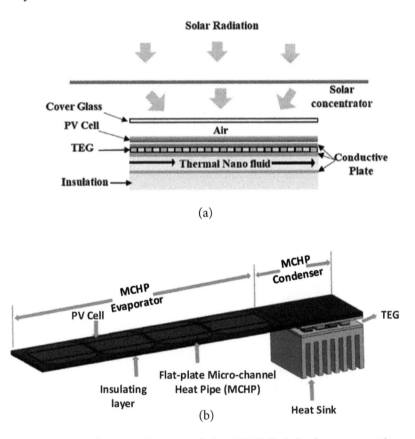

(a)

(b)

Figure 6.11 Schematic diagram of the PV-TE/T hybrid system with advanced heat exchange technology, (a) NCPV-TE system, (b) PV-TE-MCHP system [36, 39].

6.4   Spectrum Splitting PV-TE Hybrid Systems

Spectrum splitting PV-TE systems are also called tandem systems [40]. As mentioned previously, PV cells fail to convert all photons into electrical power. When the photon energy is at or above the band gap of the semiconduction material, the PV cell has a higher conversion efficiency. However, lower-energy photos not only fail to convert to electricity, but they have a negative effect on PV cell's performance due to the temperature increase of the PV cell. Therefore, the basic idea of spectrum splitting PV-TE systems is to use an optical device that makes the PV cell receive only the photons which can be used with a high efficiency, and other photons which have a strong thermal effect are directed to TEG modules and used as a heat source [5]. Figure 6.12 displays a schematic diagram of a simple spectrum splitting PV-TE hybrid system [5], which typically consists of five parts, which are the lens system, spectrum splitter, PV cell, TEG module, and heat exchange system [5]. The main function of a lens is to increase the thermal effect of radiation before the photons arrive at the hot side of the TEG module, making it possible to increase the heat flux and improve the TEG performance. The spectrum splitter is used to separate the sunlight into two different parts: one for PV cells and the other for TEG modules.

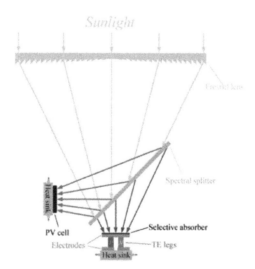

Figure 6.12 Schematic diagram of a simple spectrum splitting PV-TE hybrid system [40].

It is acknowledged that the spectrum splitter plays a dominant role in the performance of the spectrum splitting PV-TE hybrid system. For this reason, many researchers tried to propose different spectroscopic strategies or techniques in the recent decade. Firstly, the most common splitting strategy is to separate the IR or NIR light from the sunlight. In a spectrum splitting PV-TE hybrid system (as shown in Fig. 6.13) designed by Mizoshiri et al. [31], the NIR split by a hot mirror was transmitted to a TEG module, and the rest of the incident light involved UV and visible light, which was converted directly into electrical power by a PV cell. The experimental test indicated that the NIR can make the TEG module work under a temperature difference of 20°C [31]. Compared with the PV cell alone, the open circuit voltage in the hybrid system was increased by 1.3% [31].

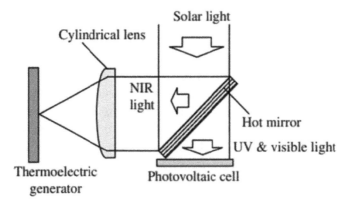

Figure 6.13 Spectrum splitting PV-TE hybrid system designed by Mizoshiri et al. [31].

Normally, the solar spectrum AM1.5D, defined by ASTM G173-03, is widely applied in the research related to PV-TE hybrid systems [41, 42]. The wavelength of the solar spectrum ranges from 280 to 4000 nm [41, 42]. According to the spectral energy distribution of a PV-TE system as shown in Fig. 6.14, there are totally different tendencies in the power fractions of PV and TEG with the wavelength cut-off increase. Therefore, in order to achieve a balanced design, the preferable cut-off wavelength is between 800 and 1000 nm for a spectrum splitting PV-TE hybrid system [28, 41, 42].

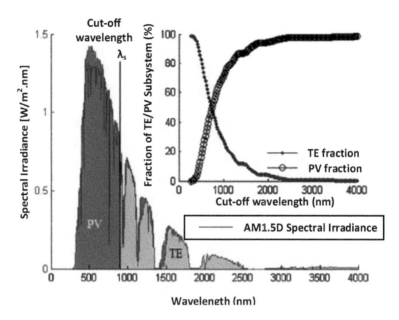

Figure 6.14 Spectral energy distribution of PV-TE hybrid system [41].

Elsarrag et al. [28] utilized a splitter made by OpticBalzers to build a PV-TE system (as shown in Fig. 6.15(a)) which can increase the output power by 10%. Figure 6.15(b) shows that the cut-off wavelength of the splitter is about 700 nm, which is very close to the preferable cut-off wavelength [28].

Additionally, some researchers indicated that coating technology can change the intrinsic property of a mirror, making it possible to create a splitter with an optimum cut-off wavelength for a spectrum splitting PV-TE hybrid system. Sibin et al. [33] utilized magnetron sputtering technology to make ITO/Ag/ITO (IAI) multilayer coatings deposit on a splitter as shown in Fig. 6.16. Through changing the thickness of the coating, they acquired a splitter for which the cut-off wavelength is 900 nm [33]. The splitter is equipped with a high reflectance for NIO-IR light (>90%) and a high transmittance for visible light (about 88%) [33].

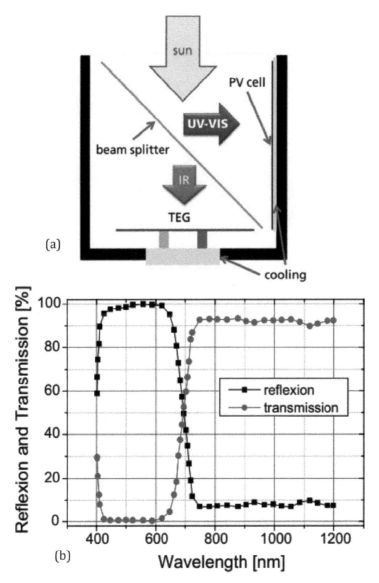

Figure 6.15 Spectrum splitting PV-TE hybrid system designed by Elsarrag et al. [28], (a) schematic diagram of the hybrid system, and (b) Transmission and reflectance data of the splitter made by OpticBalzers.

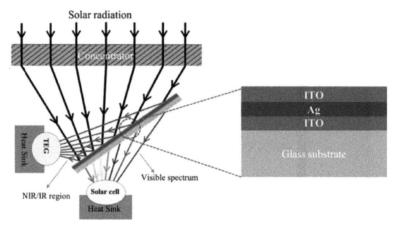

Figure 6.16 Schematic diagram of the splitter coated by ITO/Ag/ITO multilayer [33].

6.5 Conclusions

As a kind of solid-state energy convertor, TEG is considered an excellent complement to recover the wasted heat released from PV cells. Compared to the PV cell alone, the PV-TE hybrid system can improve the overall performance of the PV cell. PV-TE systems are superior to other energy recovery solutions for PV cells, such as Rankine Cycle engines and thermal storage, in that they have a compact structure and outstanding reliability. The idea of a PV-TE hybrid power system has attracted the attention of researchers since it was first proposed. This chapter provided detailed information about the working principle of PV-TE systems. The PV-TE hybrid systems can be divided into two different types: integrated systems and spectrum splitting systems. Integrated systems have a compact structure, making it possible to have excellent cost-effectiveness and reliability. Other researchers found or developed advanced cooling systems, such as micro-channel heat pipes and nanofluids, and applied them to PV-TE systems. Compared to the integrated systems, the spectrum splitting optical systems split the solar insolation into high-energy photos for the PV and low energy photos for the TEG. The spectrum splitter is complicated to manufacture; hence, it is necessary to make efforts to increase its reliability and cost-effectiveness in the future. Overall, the potential and logic of PV-TE hybrid systems are clear, but no single technology has emerged to the best in this rapidly developing field.

References

1. S. A. Kalogirou. (2014). *Solar Energy Engineering: Processes and Systems*, 2nd ed. (Elsevier: Oxford, UK).

2. M. Thirugnanasambandam, S. Iniyan, and R. Goic. (2010). A review of solar thermal technologies, *Renewable and Sustainable Energy Reviews*, vol. 14(1), pp. 312–322.

3. E. Kabir, P. Kumar, S. Kumar, A. A. Adelodun, and K. H. Kim. (2018). Solar energy: Potential and future prospects, *Renewable and Sustainable Energy Reviews*, vol. 82, pp. 894–900.

4. G. Alva, L. Liu, X. Huang, and G. Fang. (2017). Thermal energy storage materials and systems for solar energy applications, *Renewable and Sustainable Energy Reviews*, vol. 68, pp. 693–706.

5. G. Li, S. Shittu, T. M. O. Diallo, M. Yu, X. Zhao, and J. Ji. (2018). A review of solar photovoltaic-thermoelectric hybrid system for electricity generation, *Energy*, vol. 158, pp. 41–58.

6. C. Schwingshackl, M. Petitta, J. E. Wagner, G. Belluardo, D. Moser, M. Castelli, M. Zebisch, and A. Tetzlaff. (2013). Wind effect on PV module temperature: Analysis of different techniques for an accurate estimation, *Energy Procedia*, vol. 40, pp. 77–86.

7. J. K. Kaldellis, M. Kapsali, and K. A. Kavadias. (2014). Temperature and wind speed impact on the efficiency of PV installations. Experience obtained from outdoor measurements in Greece, *Renewable Energy*, vol. 66, pp. 612–624.

8. M. M. Rahman, M. Hasanuzzaman, and N. A. Rahim. (2015). Effects of various parameters on PV-module power and efficiency, *Energy Conversion and Management*, vol. 103, pp. 348–358.

9. S. Liu and M. Sakr. (2013). A comprehensive review on passive heat transfer enhancements in pipe exchangers, *Renewable and Sustainable Energy Reviews*, vol. 19, pp. 64–81.

10. K. S. Garud, J. K. Seo, M. S. Patil, Y. M. Bang, Y. D. Pyo, C. P. Cho, and M. Y. Lee. (2021). Thermal–electrical–structural performances of hot heat exchanger with different internal fins of thermoelectric generator for low power generation application, *Journal of Thermal Analysis and Calorimetry*, vol. 143(1), pp. 387–419.

11. H. Shen, H. Lee, and S. Han. (2021). Optimization and fabrication of a planar thermoelectric generator for a high-performance solar thermoelectric generator, *Current Applied Physics*, vol. 22, pp. 6–13.

12. K. Ziouche, I. Bel-Hadj, and Z. Bougrioua. (2020). Thermoelectric properties of nanostructured porous-polysilicon thin films, *Nano Energy*, vol. 80, p. 105553.

13. W. Liu, Q. Jie, H. S. Kim, and Z. Ren. (2015). Current progress and future challenges in thermoelectric power generation: From materials to devices, *Acta Materialia*, vol. 87, pp. 357–376.

14. Y. J. Wang and P. C. Hsu. (2011). An investigation on partial shading of PV modules with different connection configurations of PV cells, *Energy*, vol. 36(5), pp. 3069–3078.

15. S. Dubey, J. N. Sarvaiya, and B. Seshadri. (2013). Temperature dependent photovoltaic (PV) efficiency and its effect on PV production in the world—a review, *Energy Procedia*, vol. 33, pp. 311–321.

16. A. G. Imenes and D. R. Mills. (2004). Spectral beam splitting technology for increased conversion efficiency in solar concentrating systems: A review, *Solar Energy Materials and Solar Cells*, vol. 84(1), pp. 19–69.

17. D. Du, J. Darkwa, and G. Kokogiannakis. (2013). Thermal management systems for Photovoltaics (PV) installations: A critical review, *Solar Energy*, vol. 97, pp. 238–254.

18. S. Price, R. Margolis, G. Barbose, and J. Bartlett et al., (2010). 2008 Solar Technologies Market Report. (Lawrence Berkeley National Lab, United States).

19. Z. Zhao, S. Y. Zhang, B. Hubbard, and X. Yao. (2013). The emergence of the solar photovoltaic power industry in China, *Renewable and Sustainable Energy Reviews*, vol. 21, pp. 229–236.

20. A. Attar, H. Lee, and G. J. Snyder. (2020). Optimum load resistance for a thermoelectric generator system, *Energy Conversion and Management*, vol. 226, p. 113490.

21. P. Ponnusamy, J. de Boor, and E. Müller. (2020). Using the constant properties model for accurate performance estimation of thermoelectric generator elements, *Applied Energy*, vol. 262, p. 114587.

22. S. Fan and Y. Gao. (2019). Numerical analysis on the segmented annular thermoelectric generator for waste heat recovery, *Energy*, vol. 183, pp. 35–47.

23. L. Zhu, H. Li, S. Chen, X. Tian, X. Kang, X. Jiang, and S. Qiu. (2020). Optimization analysis of a segmented thermoelectric generator based on genetic algorithm, *Renewable Energy*, vol. 156, pp. 710–718.

24. C. Wu. (1996). Analysis of waste-heat thermoelectric power generators, *Applied Thermal Engineering*, vol. 16(1), pp. 63–69.

25. L. Chen, J. Li, F. Sun, and C. Wu. (2005). Performance optimization of a two-stage semiconductor thermoelectric-generator, *Applied Energy*, vol. 82(4), pp. 300–312.

26. O. H. Ando Junior, A. L. O. Maran, and N. C. Henao. (2018). A review of the development and applications of thermoelectric microgenerators for energy harvesting, *Renewable and Sustainable Energy Reviews*, vol. 91, pp. 376–393.

27. S. Asaadi, S. Khalilarya, and S. Jafarmadar. (2019). A thermodynamic and exergoeconomic numerical study of two-stage annular thermoelectric generator, *Applied Thermal Engineering*, vol. 156, pp. 371–381.

28. E. Elsarrag, H. Pernau, J. Heuer, N. Roshan, Y. Alhorr, and K. Bartholomé. (2015). Spectrum splitting for efficient utilization of solar radiation: A novel photovoltaic–thermoelectric power generation system, *Renewables: Wind, Water, and Solar*, vol. 2(1), p. 16.

29. T. M. Tritt, H. Böttner, and L. Chen. (2008). Thermoelectrics: Direct solar thermal energy conversion, *MRS Bulletin*, vol. 33(4), pp. 366–368.

30. V. Leonov, T. Torfs, R. J. M. Vullers, and C. Van Hoof. (2010). Hybrid thermoelectric–photovoltaic generators in wireless electro-encephalography diadem and electrocardiography shirt, *Journal of Electronic Materials*, vol. 39(9), pp. 1674–1680.

31. M. Mizoshiri, M. Mikami, and K. Ozaki. (2012). Thermal–photovoltaic hybrid solar generator using thin-film thermoelectric modules, *Japanese Journal of Applied Physics*, vol. 51, p. 06FL07.

32. Y. Da, Y. Xuan, and Q. Li. (2016). From light trapping to solar energy utilization: A novel photovoltaic–thermoelectric hybrid system to fully utilize solar spectrum, *Energy*, vol. 95, pp. 200–210.

33. K. P. Sibin, N. Selvakumar, A. Kumar, A. Dey, N. Sridhara, H. D. Shashikala, A. K. Sharma, and H. C. Barshilia. (2017). Design and development of ITO/Ag/ITO spectral beam splitter coating for photovoltaic-thermoelectric hybrid systems, *Solar Energy*, vol. 141, pp. 118–126.

34. M. Mohsenzadeh, M. B. Shafii, and H. Jafari mosleh. (2017). A novel concentrating photovoltaic/thermal solar system combined with thermoelectric module in an integrated design, *Renewable Energy*, vol. 113, pp. 822–834.

35. O. Farhangian Marandi, M. Ameri, and B. Adelshahian. (2018). The experimental investigation of a hybrid photovoltaic-thermoelectric power generator solar cavity-receiver, *Solar Energy*, vol. 161, pp. 38–46.

36. G. Li, S. Shittu, K. Zhou, X. Zhao, and X. Ma. (2019). Preliminary experiment on a novel photovoltaic-thermoelectric system in summer, *Energy*, vol. 188, p. 116041.

37. Y. P. Zhou, M. J. Li, Y. H. Hu, and T. Ma. (2020). Design and experimental investigation of a novel full solar spectrum utilization system, *Applied Energy*, vol. 260, p. 114258.

38. Q. Zhao, H. Zhang, Z. Hu, and S. Hou. (2020). Achieving a broad-spectrum photovoltaic system by hybridizing a two-stage thermoelectric generator, *Energy Conversion and Management*, vol. 211, p. 112778.

39. A. Lekbir, S. Hassani, M. R. Ab Ghani, C. K. Gan, S. Mekhilef, and R. Saidur. (2018). Improved energy conversion performance of a novel design of concentrated photovoltaic system combined with thermoelectric generator with advance cooling system, *Energy Conversion and Management*, vol. 177, pp. 19–29.

40. R. Bjørk and K. K. Nielsen. (2018). The maximum theoretical performance of unconcentrated solar photovoltaic and thermo-electric generator systems, *Energy Conversion and Management*, vol. 156, pp. 264–268.

41. X. Ju, Z. Wang, G. Flamant, P. Li, and W. Zhao. (2012). Numerical analysis and optimization of a spectrum splitting concentration photovoltaic–thermoelectric hybrid system, *Solar Energy*, vol. 86(6), pp. 1941–1954.

42. E. Yin, Q. Li, and Y. Xuan. (2018). A novel optimal design method for concentration spectrum splitting photovoltaic–thermoelectric hybrid system, *Energy*, vol. 163, pp. 519–532.

Chapter 7

Low-Risk Engineering Adaptation Strategies to Climate Change Impacts at Individual Level in Urban Areas: A Developing Country's Viewpoint

Ariva Sugandi Permana and Arthit Petchsasithon

Department of Civil Engineering, Faculty of Engineering,
King Mongkut's Institute of Technology Ladkrabang,
Bangkok, Thailand

ariva.pe@kmitl.ac.th

One of the most noticeable impacts of climate change on urban areas is the greater vulnerability of cities to urban floods due to global sea rise, and extreme rainfall depth. While mitigation strategies could not be comprehensively completed in a short time, the adaptation strategies could be, therefore, undertaken to complement the overall strategies to minimize the impacts. The study aims at proposing a viable adaptation strategy that focuses on low-risk engineering solutions, taking advantage of the high implementability of adaptation strategies by individual citizens. We discuss "living with the flood" as a response to the impacts of climate change, which has already been taking place. The response

Climate Change and Pragmatic Engineering Mitigation
Edited by Jacqueline A. Stagner and David S.-K. Ting
Copyright © 2022 Jenny Stanford Publishing Pte. Ltd.
ISBN 978-981-4877-97-8 (Hardcover), 978-1-003-25658-8 (eBook)
www.jennystanford.com

is based on a low-risk engineering approach that can be done at the individual and local levels. It is low risk and viable from the perspective of sustainability for three reasons: (1) the responses are based on locally available technology and materials (2) the ability for implementation by individuals at the local level, and it is, therefore, sustainable (3) it is based on low-impact development strategies that carry harmless to the environment.

7.1 Introduction

Impacts of climate change may vary greatly and encompass almost all aspects of human life, on water, ecology, economy, energy, transport, agriculture, and health. Because of this vast coverage, the actions to minimize the impacts must be done in synergistic ways on every front to gain substantial reductions on the impacts, through both adaptation and mitigation strategies. The Climate change adaptation has been becoming an indistinguishable conventional approach as one of the climate change strategic actions to minimize the impacts [1–3]. It has also been well known that the impacts of climate change in urban areas are, among other, floods or droughts [4], increase sea level as a result of melting polar ice caps, increase rainfall depth and intensity, or more frequent drought events [5]. The climate change impacts in urban areas would be more severe and would noticeably generate more losses in comparison to the same incident in rural areas. Therefore, the climate change adaptation strategies in urban areas are significantly important.

While the mitigation strategies of climate change are aimed at attacking the causes directly, and preventing the causes to take place, and therefore minimizing the possible negative impacts of climate change, the adaptation strategies intend to modify the people's behaviors and responses to curtail the adverse impacts that might take place in the future. The adjustment of the behaviors of the citizens, as an individual, and communities towards minimizing the negative impacts and losses of climate change would likely improve the resilience of communities on the negative impacts of climate change.

The most visible impacts of climate change in urban areas are those which are associated with environmental disasters, particularly urban floods. The long-term global changes in climatic conditions may generate the presence of extreme weather patterns that causing floods or drought at the local level including urban areas. The losses, in terms of property losses and loss of life, could be more significant in the city as denser populations, as well as countless facilities and amenities, usually exist in the city. For a city with limitless funding sources and technological know-how, the options of response to deal with the negative impacts of climate change would be abundant, and it is therefore no longer interesting to be discussed. On the other side, in the developing cities and countries with poor citizens and communities, the options are almost slim due to unlimited constraints, the challenges become tougher and stronger on many fronts, and multi-pronged approaches, along with employing available resources, must be carried out. On top of that, the most important aspect is the sustainability of the program must be maintained, in which individual citizens should be able to participate with their capacity and resources. The overall planning and coordination, however, must be at the hand of the local authority.

In the view of the adaptation strategies to climate change in urban areas associated with environmental disaster, there are four essential elements of the city, which are closely associated with the climate change adaptation strategy. They are urban planning, land use planning, water resources management, and disaster management. These interlocking elements could be used as the fundamental platform in minimizing the negative impacts of climate change in urban areas with urban society as the core of the strategies. Here, society comes into the picture because of a practical purpose where sustainability is deemed important. Sustainability is not possible without the involvement of society. In the future, society must be able to run the program on its own without the necessity of deep involvement of the local authority. In this situation, the adaptation program will be perpetual and independent of the immersion of the authority financially. The interlocking of urban planning, land use planning, water resources management, and disaster management as a platform of climate change adaptation strategy is illustrated in Fig. 7.1.

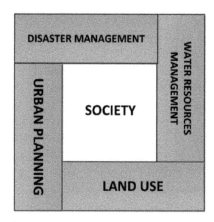

Figure 7.1 Four Urban elements in adaptation strategy with society as core.

Based on our observation, and a study was done by [6–7], one of the impacts of climate change suffered by most urban areas is a frequent flood. The responses to coping with this episode could be in several forms, for instance, mitigate the causes of climate change and flood, or if mitigation strategy is not possible, adjust our life to the flood characteristics [8–9]. Looking at climate change as the cause of the flood, the flood is more a natural phenomenon than a natural disaster. Flooding becomes disastrous when it brings about harm, such as life and property losses, damage to and destruction of public infrastructure, disruptions to the socioeconomic functioning, and provision of services to society. The flooding problem also brings other negative environmental effects with immediate and long-term social, ecological, and economic implications [10]. It happens because human society has put itself at risk by settling down in harm's way, carrying out agriculture, industry, and economic activities and other activities such as constructing roads, bridges, and railway lines in areas susceptible to flooding [11]. In densely populated and rapidly developing land-scarce urban areas, the magnitude of the disaster caused by flooding understandably swells as more people and development activities encroach onto floodplains. The activities also bring about the worst impacts when adequate mitigation and adaptation strategies are absent at either the institutional or the grassroots level to cope with floods.

7.2 Mitigation vis-à-vis Adaptation Strategies

As far as we are concerned, mitigation strategies anticipate the reduction or limiting greenhouse gas emissions, and the adaptation strategies heighten the susceptibility to the effects of climate change. As [12] highlighted some critical points in implementing the adaptation strategies:

- Where vulnerability in a community is high and where the need for safety and resilience is urgent, the adaptation efforts in communities must therefore be prioritized.
- The adaptation strategies must be streamlined into long-term national and local sustainable development and poverty reduction strategies.
- In proposing climate change adaptation strategies, the projected climate change-related trends on risk and vulnerability assessment must be based on sufficiently long data of climate variability.
- The strategies must prioritize the strengthening of existing capacities among local authorities, civil society organizations, and the private sector. This is to lay the foundations for the robust management of climate risk and the rapid scaling up of adaptation through community-based risk reduction and effective local governance.
- To ensure the flow of both financial and technical supports to local actors, a strong resource mobilization to support the adaptation strategies must be developed.
- Improved early warning systems, contingency planning, and integrated response must be in place to support effective community-based adaptation and risk reduction.

The above points need further elaboration to work properly for the implementation at the local and individual level. Notably, these adaptation strategies may not be sufficiently operational without the complement of mitigation strategies. The mitigation strategy associated with urban elements in urban areas includes:

- Road and network arrangement: This is intended to encouraging public transport, ride-sharing, discouraging private transport, and "within walking distance" principles, and therefore reducing the greenhouse gas emissions in the city.

- Building aspect: This is aimed at promoting energy-efficient buildings, promoting green buildings, and low-impact building materials, and therefore reducing energy uses of the buildings.
- Open space and greenery: It is promoted to increase urban forestry to maximize the capacity of urban sequestration, and thus increasing the capacity of air pollution absorption.
- Reducing the causes of urban heat island, and expanding the urban lakes and other water bodies to modify the micro-climate in the city.
- On the housing aspects: Housings are the most predominant building in the urban area, therefore tackling the climate change issues associated with this aspect would significantly reduce the greenhouse gas emissions in the city. The actions necessary are by promoting sustainable housing, increasing the capacity of rainwater harvesting, reducing the quantity of surface runoff, avoiding flood inundation by elevating the floor, and other flood-proofing interventions as part of adaptation effort to climate change impacts.
- Zoning is essential as a passive strategy to avoid disaster, a zonation strategy can be applied by considering the vulnerability of a particular part of the city to disasters associated with climate change.
- Urban ecosystem: what we can do to conserve the ecosystem of the city is, among others, by minimizing the disturbance to the natural ecosystem that exists within the city and periphery.

Among the above adaptation and mitigation strategies, our discussion is focused on the viable climate change adaptation strategies at local and individual levels from the perspective of developing cities. The viability of the individual level is selected with the reason for sustainability. When the strategies can be implemented sustainably then the strategies will be viable from social and economic viewpoints. Individuals are the lowest stratum in the hierarchy within the community, when a plan is implemented and it works and is maintained in a long run at the individual level independently, this means the plan is sustainable. Based on the experiences of projects in developing countries, financial independence from the government's hands out is one of the most important factors of sustainability. The conceptual framework of the viable climate change adaptation strategies at individual and local levels is shown in Fig. 7.2. This issue is the focus of this study. At this point, our

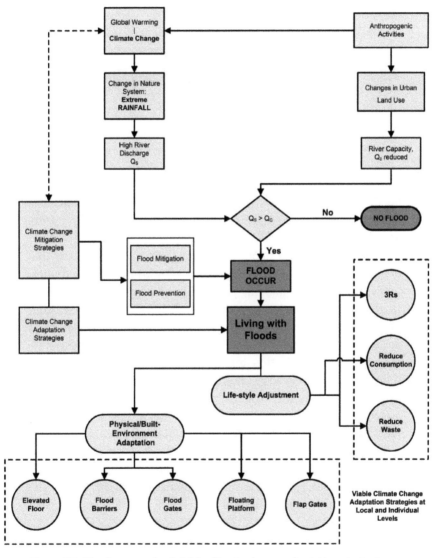

Figure 7.2 The framework of viable climate change adaptation strategies.

focus of discussion on "living with the flood" is emphasized on physical/built-environment adaptation, rather than lifestyle adjustment that may include waste and consumption matters such as adopting the culture of reducing, reusing, and recycling (3Rs) in waste management at individual and community levels; reducing consumption, and reducing wastes. Discussion on these matters would neither appeal nor have the same effectiveness as their physical/built-environment adaptation counterpart.

7.3 Climate Change as the Primary Cause of Urban Flooding

Figure 7.2 implicitly shows the causes of flood. Depending on the stage at which we start to observe the cause of the flood, the cause might be different. However, if we trace all the causes of flood along the process, we would find that the primary causes of the flood are nature and people. From nature's viewpoint, the climate change that may cause extreme weather conditions, i.e. extreme rainfall or prolonged drought can be considered the primary cause of the flood, even though it is believed that one of the causes of climate change is anthropogenic activities. In line with this, the flooding problem in urban areas is largely due to the urbanization process, where changes in land use from previously natural to built-up areas are frequent. Although urban areas occupy less than 3% of the Earth's land surfaces, the effect of urbanization on flood hydrology and flood hazard is disproportionately large (Smith and Ward, 1998).

One of the most widely used equations to calculate the quantity of discharge is the rational formula, first described by [13]. The formula is expressed by $Q = CIA$, where Q denotes the estimated maximum flood discharge (m^3/s); C, a runoff coefficient that indicates the percentage of rainfall and appears as overland flow (dimensionless); I, the rainfall intensity (mm/h); and A, the area of the watershed within which rainwater that falls will flow through a certain reference point (hectare).

The runoff coefficient is dependent on the type of surfaces, for example, asphalt paving in good order will have a runoff coefficient as much as 0.85 to 0.90, and this parameter indicates that 85% to 90% of rainfall will be transformed into an overland flow that may lead to flooding. Unlike asphalt paving, parks, gardens, and lawns (depending on the slope of its surface and character of sub-soil) have runoff coefficient of 0.05 to 0.25. These two conditions show that in a more urbanized area where the built-environment is dominant, the same rainfall intensity will generate much greater storm discharge compared to an area that is predominantly a natural environment. That is why in the urban area where inadequate or poorly maintained drainage channels are present, flooding tends to occur more frequently.

Limited land and other resources are the most common problem in urban areas. Urban poor communities tend to suffer the most in any urban problems, including having limited choices of suitable urban land for their settlements; some of the poor communities thus end up inhabiting the floodplain that is periodically flooded. In reality, it is not only the poor communities who reside on floodplains but also other strata of the communities. One solution to this problem is the introduction of "living with the floods" as an alternative to alleviate flood risks that arise from the necessity to occupy floodplains due to land scarcity in urban areas. Living with the flood requires a certain acceptable condition of the floods in terms of safety for the individual and community. Two important flood parameters significantly affect people's response to flood: depth (D) and velocity of the flood (V). These two parameters will determine whether living with a flood can be applied without compromising the safety of the community and will thus be discussed in the following sections.

7.4   Floodwater Depth-Velocity Correlation

The degree to which a flood poses hazards upon human beings hinges on two flood parameters: floodwater depth and velocity. The combined effects of these parameters help suggest the most suitable solution to flooding problems in any type of situation, including one that employs flood-proofing activities.

Figure 7.3 shows the flood risks on human activities according to flood velocity and depth. Depending on these two variables, generally, the flood risks on human activities can be categorized into three, namely safe zone, potentially unsafe zone, and dangerous zone. The flood situation is categorized as safe for healthy adult people when the zone is within the less than 150 cm depth of stagnant water, and maximum floodwater velocity of 80 cm/s for a floodwater depth of less than 30 cm. The flood can be dangerous for a human being if the flood velocity is greater than 125 cm/s with a minimum depth of 30 cm because, at this point, an adult people could probably be swept out. At the other end of the danger zone is when the flood depth is greater than 180 cm even with zero velocity since a non-swimmer adult person will be drowned. The other combination of the two variables will define the risks.

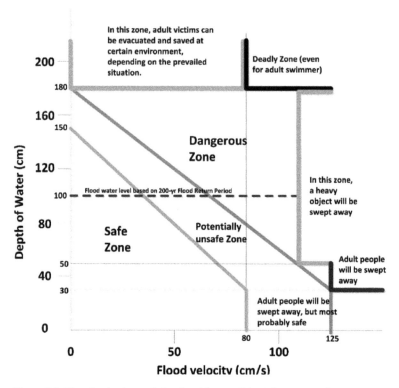

Figure 7.3 Flood velocity and depth with possible risks on people.

Before discussing the details of "living with the floods" as a generic theme of climate change adaptation strategies at the individual level, a theoretical excursion is necessary to provide background information on conventional flood prevention efforts. These efforts are universally undertaken and have been proven effective to some degree. The elaboration of the climate change adaptation at the individual level can proceed from this point.

7.5 Flood Prevention

Flood prevention is a general method of avoiding excessive negative impacts due to flood before it happens. It may involve structural measures such as the construction of levees and reservoirs, or non-structural methods such as land-use planning, floodplain management, early warning system, and the likes. Both structural and non-structural measures must complement each other to minimize or, if possible, eliminate losses due to flooding. It emphasizes "losses" since the occurrence of the flood cannot be completely averted. The following sub-sections briefly discuss those methods.

7.6 Structural Measures

This method is achieved by undertaking engineering construction such as improvement of channel capacity, installation of levees, floodwalls, storage dams, retention basins, and the likes. The first three are aimed at confining stormwater within designated channels. Meanwhile, storage dams and retention basins seek to control and regulate floods by releasing water to the downstream gradually so that river channels at the downstream reaches are capable of safely conveying the stormwater without overflowing. Figure 7.4 shows a schematic representation of structural measures using the improvement of the channel capacity and construction of flood protection dikes. Flood channel or floodway is improved to be able to confine stormwater adequately in such a way that does not create overflow. To confine stormwater within the designated channel section, the channel capacity should be sufficient to accommodate the stormwater.

To plan and design an adequate channel capacity, the term of *return period* is introduced. The return period of a flood is defined as the probability of a certain quantity of discharge being equaled or exceeded once in every given period (years). All flood control facilities are designed based on certain return periods; in Thailand, for example, return periods of 25 to 50 years are predominant, while Japan applies flood return periods of 10 to 200 years [14].

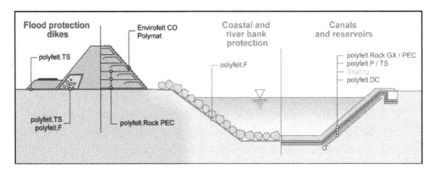

Figure 7.4 Improvement of channel capacity.

Another means of structural flood prevention is the retention basin (refer to Fig. 7.5). As the name implies, the retention basin is aimed at reducing flood discharge at its downstream by diverting part of the flood discharge into the basin (Q_{in}). The basin subsequently releases the floodwater gradually into the downstream (Q_{out}) in such a way that the capacity of the river in downstream reaches should be able to accommodate the regulated discharge released from the upstream.

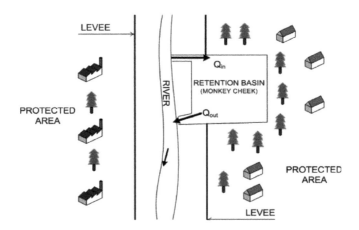

Figure 7.5 Retention basin.

By those arrangements, the protected area along the river particularly the downstream of the retention basin is prevented from being flooded. The capacity of retention basins should be adequate to accommodate flood design while considering river capacity at the downstream reach of the basin. The main objection to this system is that it needs large parcels of land. It is because, in many cases, retention basins are located on flatlands and the main purpose of the basins is to control as large a volume of floodwater as possible. It becomes too luxurious when only limited land resources are available in the urban area. Another limitation of the system is that it only protects areas downstream of the basin. Notwithstanding the above, the fringe benefits of retention basins are possible utilization of the water contained within the basin for water supply and the possibilities of recreation and eco-tourism attractions.

The construction of a dam shares essentially the same principles as that of the retention basin in flood prevention. Nonetheless, a dam is much more complicated in planning, design, construction, operation, and maintenance compared with a retention basin. The social and environmental impacts of a dam are also more significant than that of a retention basin. However, the volume of floodwater that can be controlled by a dam is significantly larger than that of a retention basin due to the greater size and height of the dam.

Structural measures are usually costly, and the possible impacts on the environment and society associated with the measures are also great; in some cases, it is more undesirable than other flood prevention systems. Consequently, other alternatives that are less costly and with smaller environmental impacts have been sought after.

7.7   Non-Structural or Regulatory Measures

Non-structural measures can be undertaken in various ways such as land use planning, floodplain management, flood risks mapping, early warning system, and flood-proofing. Non-structural measures normally require legislative backing to achieve their objectives. Non-structural measures alone will not be effective and work best when combined with other flood prevention efforts. The main advantage of non-structural measures is that they can be carried out at the community level and community participation is as such largely accommodated. The measures are briefly described below.

7.7.1 Land Use Planning

Concerning flood prevention, urban land use planning has two objectives: first, reducing losses due to floods by avoiding occupation on flood-prone lands, and second, reducing the quantity of flood discharge by decreasing the composite runoff coefficient (C) of the urban land with more natural surfaces (refer to equation $Q = CIA$). More recent ideas include the introduction of low impact development (LID) and "Green Streets" that require the installation of rainwater harvesting mechanisms, rain gardens, bio-retention swales, pervious paving for parking areas that all aim at reducing the urban surface runoff during storm events.

7.7.2 Flood Plain Management

Flood plain management aims at achieving the compatibility of human activities with a flood when these activities necessarily take place on land designated as a floodplain. All designated floodplains are inventoried and assessed according to their risks due to floods; this process may employ the $V–D$ correlation as shown in Fig. 7.3.

7.7.3 Flood Risks Mapping

Flood risk maps show possible losses for different degrees of flooding. By assessing flood risk maps, therefore, losses may be predicted for a forecast flood event. Employment of the GIS technology has enhanced the efficacy of flood risks area delineation and prediction of flood losses. When a flood risks map is superimposed over infrastructure and socio-economic maps, possible flood damage and losses can be accurately visualized. Without being implemented appropriately through land use planning practices, flood risk mapping itself does nothing to prevent flood loss. Legal support to bring flood risk mapping objectives into practice is required. Consistent implementation and public participation are, therefore, key elements to the success of goal achievement of flood risks mapping activities.

7.7.4   Flood Early Warning System

An early warning system of the flood will help to minimize damage and losses. Adequate warning lead time will help to minimize the damage and losses significantly. Flood early warning system is expected to work simultaneously with other flood mitigation efforts in a coordinated manner.

Warning lead time becomes highly crucial because the minimization of damage and losses depend largely on whether sufficient warning lead time has been given to potential victims to take alleviation actions. The generic correlation between the depth of floodwater and gross damage and losses potentially suffered under varying warning lead times. It seems obvious that damage and losses would be significantly reduced if early warning could be issued to potential flood victims. However, huge losses were still suffered in flood events as recent as 2009 due to the absence of early warning. For instance, flood victims in Central Vietnam due to Typhoon Mirinae cited the absence of official warning and the resultant lack of time to respond as the first of three main reasons for their losses [15].

7.7.5   Flood-Proofing: The Basic Concept of "Living with the Flood"

Flood-proofing can be undertaken at the individual level within the community; it is an example of the best-possible adaptation of humans to the natural phenomenon, and this is the essence of living with the flood.

As one of the significant impacts of climate change, dealing with the urban flood is urgent as the losses are usually substantial. The impacts of the urban flood can be minimized by various efforts as discussed above. However, when the flood itself cannot be minimized due to the increasing pressure of climate change coupled with uncontrolled human activities, the adjustment will be necessary and unavoidable through, for example, "living with the flood". Living with the flood is an attitudinal fine-tuning that can be done at the individual community simultaneously and sustainably, and it is, therefore, viable from the socio-economic viewpoint.

7.8 Living with the Flood as Climate Change Adaptation Strategies at Community Level

Land resources in the city are limited. In many cases, flood-prone areas are encroached upon and occupied for habitation and various urban functions. To remove the human occupation on flood-prone lands would not be realistic from both social and economic viewpoints. The conflict between humans living on floodplains and periodic flooding needs to be resolved through appropriate physical adaptations to the living environments. This is to accommodate the occasional intrusion of water while disruptions to people's lives are kept to a minimum. These will also be realistic and low-risk climate change adaptation strategies in urban areas despite the complexity of the issues in the city.

7.8.1 Flood-Proofing

Flood-proofing is defined as any combination of techniques used to change the structure or property to reduce or eliminate flood damage. The techniques of flood-proofing include berms, floodwalls, closures or sealants, elevation or relocation, and any other techniques that, in principle, offer on-site flood damage protection.

7.8.2 Elevated Floor

There are two methods to elevate a house's floor, by using piles and land-filling. Land-filling is undesirable because the volume taken up by land-filling must be compensated for with additional water depth since the method just shifts flooding problems somewhere else. The pile's system does not require depth compensation, since the volume replaced by this system is negligible (refer to Fig. 7.6). Access to the home is achieved by constructing a ramp from the nearest ground surface which has a higher elevation than the expected flood level.

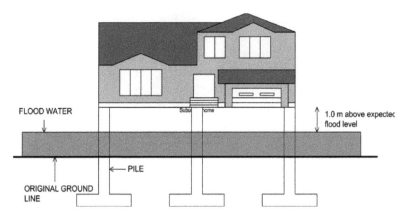

Figure 7.6 Elevated floor by piling system. Adapted from [16].

Another system of an elevated floor is by employing land-filling (see Fig. 7.7). If this system is used extensively within a certain area, it will shift the flood problem somewhere. Alternatively, else, it just reallocates but does not solve the problem locally. Because of this disadvantage, this system is undesirable compared with the piling system. If buildings have already been there before flooding problems occur, to modify the building by elevating its floor will be quite costly. In this case, flood-proofing may be done by constructing floodwater barriers encircling the property as shown in Fig. 7.8.

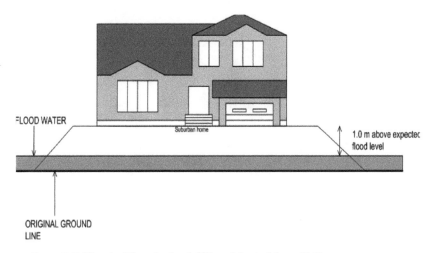

Figure 7.7 Elevated floor by land-filling. Adapted from [16].

Figure 7.8 Individual flood barriers. Adapted from [16].

Flood barriers can be constructed from concrete floodwall or soil dikes encircling the property. Access to the home can be made by using sealed and waterproof gates at the barrier. However, during floods, the opening gate should be properly closed and sealed. The disadvantage of this method is similar to the land-filling system since it displaces some amount of floodwater volume and shifts it somewhere else. If the system is widely used by individuals the flooding problems just shift to other adjacent locations, and it perhaps becomes economically more efficient to utilize the ring-bound system that is a dike encircling the community as a whole as shown in Fig. 7.9.

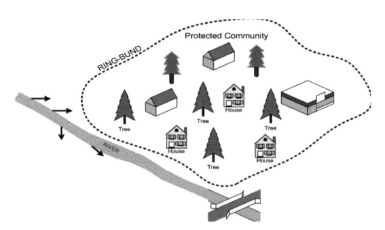

Figure 7.9 Ring-bund encircling the community.

If the initiative of ring-bund construction and most of the development costs come from the community itself, the effort can still be categorized as flood-proofing at the community level. However, if the construction covers the community at large and the local authority bears the development costs, the effort is then considered part of the structural measures, which will be a costly initiative.

If the depth of floodwater is less than 30 cm with a maximum flow velocity of 0.5 m/s, the individual homes can do flood-proofing efforts by using waterproof or high impermeability, elevated concrete foundations such as shown in Fig. 7.10. Flood flow of less than 0.5 m/s is considered as non-erodible overland flow.

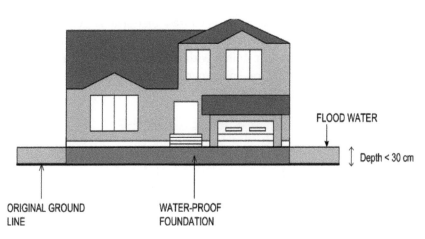

Figure 7.10 Home with waterproof foundation.

With a floodwater depth of less than 30 cm and foundations made from waterproof material, structurally to the house, the duration of the flood does not become a problem as long as the homeowner's activities are not adversely affected by the situation. If the expected depth of flood is higher but still acceptable in terms of safety, a floating platform can be considered for the part of the house, e.g., a garage or kitchen, or even the entire house (a floating home) if the size of the house is relatively small. It will be further discussed in the following sub-section.

7.8.3 Floating Platform (Garage, Kitchen, Pathway)

According to Archimedes' law, a body submerged in fluid experiences a buoyant force equal to the weight of the displaced fluid. If this law is applied to the flood-proofing system, for a floating platform to lift a garage, if the total weight of the garage is 4.0 tons, therefore, the volume of the floating apparatus must be at least 6.0 m^3. Considering the cost element, the maximum lifting capacity of the platform must be minimized. With a minimum capacity of the platform and minimum size of the floating apparatus, the only appropriate utilization is perhaps for the garage or kitchen. These floating premises (refer to Fig. 7.11) will work appropriately at relatively deep floodwater but with quite a slow velocity, it is desirable if the velocity of blood flow is zero during flood occurrence. The anchored platform is required to avoid movement due to flood flow; in this case, the platform is anchored at least at four points with guard-rail to four firm walls or piles, to allow vertical movements of the platform following rising and falling floodwaters.

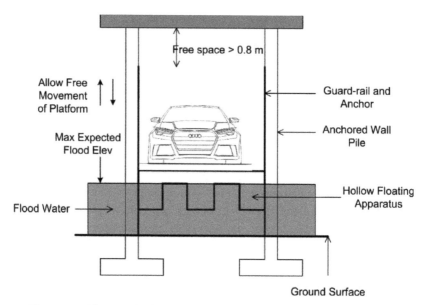

Figure 7.11 Floating platform for garage.

This idea follows from the condition that there are no feasible alternatives to avoid flood events; living with the floods, therefore, becomes necessary. Floodwater contains latent energy to be utilized; the buoyancy force of floodwater is utilized in this particular matter to minimize property loss and disruption due to flood. This method does not require high technology to be implemented; an appropriate level of technology that is easily available in the locality will work. However, experimental work to prove the effectiveness of the system must be undertaken before implementation since this idea is perhaps introduced.

The initial volume of fluid, e.g., floodwater displaced by the hollow floating apparatus, must be large enough to initially lift the platform. Periodical checking on the hollow floating apparatus must be undertaken to ensure that there is no leakage on the apparatus. A pre-fabricated light reinforced concrete, such as Ferro-cement, is perhaps good for hollow floating apparatus. The minimum volume of the floating apparatus can be estimated by using the formula of V_{min} [in m^3] $= 1.5 \times \dfrac{\text{Total Weight [in kg]}}{997}$ where total weight is the weight of the load and the self-weight of the floating apparatus.

7.8.4 Flood Barriers or Flood Guards

This flood-proofing effort emphasizes on-plot barriers such as temporary door or window closure by using pre-fabricated closure equipment or temporary sandbags (refer to Fig. 7.12).

Pre-fabricated flood barriers are commercially available. However, a cheaper temporary flood barrier can be produced locally by using polyepropylene bags filled with locally available clay soil. The barriers can also use gunny sack, also known as a gunny shoe or tow sack, is an inexpensive bag, traditionally made of hessian fabric formed from jute, hemp, or other natural fibers. The size can be varies up to 74 cm (width) × 23 cm (depth) × 110 cm (length). If the bag is 90% filled up with the sands, the weight would be around 80–90 kilograms. The working principle of sandbags in preventing floodwater is very simple; it provides the height flexibility, the strength of the stack of bags increases by its own weight under gravity.

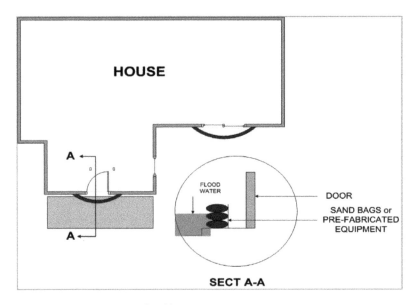

Figure 7.12 Temporary flood barriers.

7.8.5 Floodgates

By "floodgates," it is here meant a pre-fabricated "gate" that is installed as the closure of the front-gate opening of the fence. This effort assumes that there is an encircling flood-proof fence that is sufficient to prevent the flood. The term "sufficient" here is concerning its height, construction, and material. Since it is pre-fabricated equipment, it is not a cheap flood-proofing tool, and for certain people, it will not be affordable. It is shown in Fig. 7.13.

The installment of floodgates will only be effective if the front gate of the encircling fence is the only opening available surrounding the house, and again providing that the fence is sufficiently reinforced to withstand horizontal hydrostatic pressure that is due to flooding depth and velocity. Please note that floodwater leakage may occur at the boundary between the gate and the wall or hinges. Therefore, during the floodgate installation, careful attention must be given to these parts.

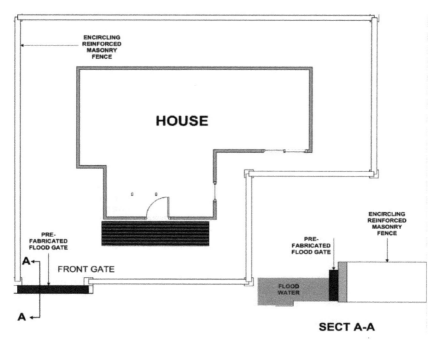

Figure 7.13 Installment of floodgate.

7.8.6   Flap Gates and Backflow Valves

House drainage is perhaps the most open system that connects flood flows outside the house into the house. The wastewater/sewerage system may leak if it uses the closed pipe system while the clean water supply is less prone to leaking compared with the other two. Flap gates can be utilized to prevent floodwater seepage from the inside part of the house if the house drainage system uses the open channel. This system works by utilizing the differential hydrostatic pressure between the outside and the inside; this principle can be depicted in Fig. 7.14. Hydrostatic pressure P (ton/m^2) is computed according to $P = \frac{1}{2}\, w{*}h^2$ where w is the density of fluid, e.g., floodwater (ton/m^3), and h is the depth of water (m).

Flood-proofing does not reduce the flood quantity. There are community-based activities that can significantly reduce the flood magnitude, this action employs a similar principle with the retention basin or flood control dam, but it is implemented at the community level. The following section discusses this matter.

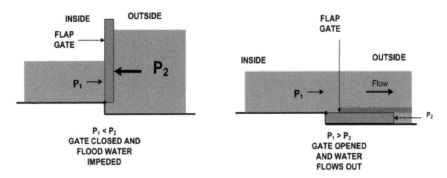

Figure 7.14 Automatic flap gate.

7.8.7 Reducing Runoff and Rainwater Harvesting

Reducing runoff by using this technique would do also rainwater harvesting, and to some extent, dealing with drought simultaneously. Floods are greatly influenced by urban land use. Not many local authorities can appropriately control land use to reduce flood magnitude. Given the present runoff coefficients are not easy to modify and also rainfall is beyond people's control. Thus, the maximum discharge from an urban area theoretically cannot be modified. However, the rate of release of discharge can be regulated through concerted actions from the community; how does it work? It acts like a distributed storage system; it can be undertaken if all individuals within the community are willing to contribute to reducing the flood magnitude.

All individuals are asked to provide storage capacity, and to ensure fairness in distribution, the storage which is provided by each in the community should be based on the area of individual land plots. The local authority, in this case, determines the design rainfall that will be regulated by the decentralized system. For example, a design rainfall is designated as h mm/h. The individuals, therefore, provide storage according to their land plot area, defined by $S_i = 0.001A_i*h*D$, where S_i is storage that must be provided individually (m^3), A_i is individual land plots area (m²), *h* is design rainfall (mm/h) determined by the local authority, and *D* is projected rainfall duration (h). The best situation will be created if those storages are installed underground since this enables collected rainfall to recharge into groundwater. In the long run, it will provide sufficient groundwater sources and ultimately lead to sustainable development (refer to Fig. 7.15). Rainwater that falls within an individual land parcel is collected, including through pipes from the rooftop, and discharged into an underground tank for subsequent recharge into groundwater.

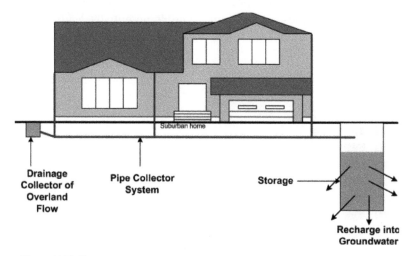

Figure 7.15 Storage systems.

The effectiveness of the individual storage system depends on the hydraulic conductivity of tank storage, groundwater table, as well as rainfall intensity, duration, and frequency. More permeable soil structure around the storage boosts groundwater recharge, therefore, the process of emptying the storage will be faster, and successive rainfall can be stored properly in the tank. Higher groundwater table and less permeable soil structure will delay the emptying process of the tank and reduces its capacity for storing successive rainfall.

In a densely populated urban area, where detached individual houses are normally rare, and multi-story building types are dominant, the storage system can be placed at either rooftop or basement (refer to Fig. 7.16). However, a different operation is applied for the rooftop storage, that is, at the time when rainfall stops, and underground storage is empty, the rooftop storage can then be released to the underground storage. The same principle of storage calculation for individual detached houses can be applied to multi-story buildings. With this arrangement, assuming that the individual storage system works well, the reduction of flood magnitude will be directly proportionate to the built-up area excluding roads and other non-occupancy areas. This reduction also leads to a reduction in the need for drainage infrastructure; costs for providing such infrastructure; and flood damage and losses. At the same time, it potentially leads to an increase in groundwater resources and improved environmental sustainability.

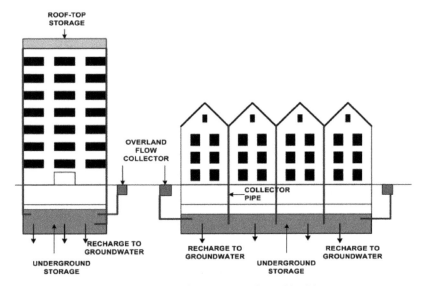

Figure 7.16 Storage system in urban-type residential building.

If all the above-mentioned activities are implemented appropriately, harmonious coexistence between humans and nature will be potentially achieved. Floods will no longer be viewed as disasters to defend against, but rather as normal, natural phenomena that humans must learn to adapt to and make the best of. The Netherlands' socially-rooted approach for addressing climate change adaptation to flooding sums it up best with a vision of a country "safe against flooding, while remaining an attractive place to live, to reside and work, for recreation and investment" [17].

7.8.8 Recharge Wells

The objective of the provision of the recharge well is to reduce the overland flow during raining, by transferring overland flow into good storage. Since the recharge well is provided at individual land plots, therefore the recharge well can also be provided by an individual with an impartial basis, as the larger the land plot area, the larger the landowner must provide the recharge well. The recharge well can solve two water-associated disasters simultaneously: the flood and the drought. While the flood is solved by reducing overland flow, the drought is eliminated by increasing groundwater storage in the local aquifers.

A recharge well may only work sound if the hydraulic conductivity of the soil at the location where the recharge well is constructed is high, for example, sandy soil. However, this issue is only about when the vacating time of the well-storage matters. The vacating time is the time required to empty the recharge well storage, and the well will therefore be able to refill at the next cycle of rain. The basic design calculation of recharge well is as follows (Fig. 7.17).

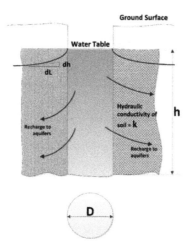

Figure 7.17 Recharge well.

The volume of recharge well storage can be determined from the land plots owned by the individuals, and based on the design rainfall designated by the authority. If the land plot area is A m^2, and the design rainfall is r m, the volume of the recharge well is, at least, $V_r = A \times r$ [m^3]. The vacating time of the recharge well to be able to receive rainfall in the next cycle of rain can be determined by the following formula:

$$t = \frac{V_w}{\pi Dhk \dfrac{\Delta h}{\Delta L}} = \frac{0.25\pi D^2 h}{\pi Dhk \dfrac{\Delta h}{\Delta L}} = \frac{D}{4k \dfrac{\Delta h}{\Delta L}},$$

where D is the average diameter of the recharge well [m], h the depth of water in the recharge well [m], k the hydraulic conductivity of the soil [m/day], $\dfrac{\Delta h}{\Delta L}$ the hydraulic gradient of the groundwater table [non-dimensional], V_w the volume of water in the recharge well [m^3], and t total vacating time of the storage [day].

The volume of the recharge well must be slightly larger than the volume of water to be stored in the well. This is to avoid overflow from the well.

The advantages and disadvantages and the basic requirements of the different adaptation approach through "living with the flood" can be summarized in Table 7.1.

Table 7.1 Basic requirements, advantages, and disadvantages of the approach of "living with the flood"

No.	The approach	Basic requirements	Primary advantages	Primary disadvantages
1	Elevated floor by piling system – Fig. 7.6	Good soil for pile foundation in the site.	It does not substitute the floodwater. Thus it does not shift the floodwater elsewhere.	The construction cost may probably high.
2	Elevated floor by land-filling – Fig. 7.7	Availability of sufficient soil for the embankment.	Can be constructed in poor soil condition for the foundation.	It shifts the flood elsewhere.

No.	The approach	Basic requirements	Primary advantages	Primary disadvantages
3	Individual flood-barriers – Fig. 7.8	None.	Individual flood defense system (It can be done at an individual home).	It shifts the flood elsewhere, and the only capable individual can do.
4	Community ring-bund – Fig. 7.9	Availability of sufficient soil for the embankment, and sufficient land area en-route.	Collective flood defense system.	High construction costs.
5	Flood-proof foundation home – Fig. 7.10	None.	Locally available construction materials.	It shifts the flood elsewhere.
6	Floating platform – Fig. 7.11	It needs sufficient land area.	Flexibility, and it does not shift the floodwater elsewhere.	High construction cost.
7	Temporary flood-barriers – Fig. 7.12	None.	Locally available construction materials.	None.
8	Floodgates (at encircling fence) – Fig. 7.13	The individual home must have a flood-resistant fence in place.	Individual flood defense system (It can be done at an individual home).	The basic requirement may not be readily available at an individual home.
9	Automatic flap gate – Fig. 7.14	Gate or canal is in place.	Depending on the situation, it can be done individually or collectively.	It may involve a high construction cost.
10	Recharge well from rooftop rainfall – Fig. 7.15	The hydraulic conductivity of soil in the site must be considerably high, e.g., sandy soil.	Reducing overland flow and increasing groundwater recharge at the same time.	Cannot be constructed at any place/ condition.
11	Underground storage at the urban-type residential building – Fig. 7.16	Underground storage is available.	Reducing overland flow, and to some extent, recharge to groundwater.	None.

The "living with the flood" approach as one of the adaptation strategies shown in Table 7.1 does not at all reducing the flood quantity. It is rather the way of individuals adjusting to the natural phenomenon that has already happened and will be continuously happening if no efforts have been done to reduce the flood itself through other means. It is socially and environmentally a low risk, as the implementation of any of this approach will not lead to social and environmental devastation, and therefore viable from the socio-environmental viewpoint.

7.8.9 Concluding Remarks

For the effectiveness of the implementation, climate change mitigation and adaptation strategies are expected to go hand-in-hand or complement one another. Due to the characteristics of mitigation strategies, the strategies are mostly executed with extensive financial supports, and therefore can only be implemented by the government as the main actor, even though the individual citizens can also be the actors. On the other hand, adaptation strategies can be implemented by individual citizens as the actors, or with a little thrust from the government to run. The adaptation strategies may encompass the plans or programs or projects from easy and inexpensive to sophisticated and affluent ones. It depends on the capacity of the individuals in contributing to fighting climate change and minimizing the impacts through adaptation strategies. By this feature, the citizens are free to adopt one or more strategies and every individual can contribute.

Despite its high sustainability due to implementation at the individual level, "Living with the flood" as one of the adaptation strategies of climate change has no impact to prevent climate change impacts. The efforts are just to reduce the losses of climate change impacts, and it is necessary to be accompanied by mitigation strategies. Another robust point of the "living with the flood" that makes this effort sustainable and implementable is that no sophisticated technology is involved—the simple technology available locally and the ability to employ the local materials, and certainly local labor.

The other strategy that, in our opinion, can be categorized as adaptation strategies is life-style adjustment. Examples of the activities that include in this category are 3Rs (reduce, reuse and recycle) at individual households, reduction of consumption of, for instance, energy, and reduction of waste. These are active adaptation strategies in reducing the causes of climate change. This active adaptation strategy does not require any great sources to implement; it needs only the willingness of the individuals. Even, implementing this strategy might save the source. However, we do not discuss this matter, as we need to focus on living with the flood.

Unfortunately, we cannot present the real-world examples of the issue in a complete picture. But, trying to implement this strategy in flood-vulnerable urban areas would be worthwhile, and do the monitoring afterward to understand the effectiveness of the strategy.

References

1. Burton, I., Diringer, E., and Smith, J. (2006). *Adaptation to Climate Change: International Policy Options*. Arlington: Pew Center on Global Climate Change.

2. Brunner, R., and Lynch, A. (2013). *Adaptive Governance and Climate Change*. Springer Science & Business Media.

3. Field, C. B. (ed.). (2014). *Climate Change 2014–Impacts, Adaptation and Vulnerability: Regional Aspects*. Cambridge University Press.

4. USGCRP (2014). *Climate Change Impacts in the United States: The Third National Climate Assessment.* Melillo, Jerry M., Terese (T. C.) Richmond, and Gary W. Yohe (eds.). United States Global Change Research Program.

5. Bulkeley, H. (2013). *Cities and Climate Change*. Routledge.

6. Khailani, D. K., and Perera, R. (2013). Mainstreaming disaster resilience attributes in local development plans for the adaptation to climate change induced flooding: A study based on the local plan of Shah Alam City, Malaysia. *Land Use Policy*, 30(1), 615–627.

7. Huong, H. T. L., and Pathirana, A. (2013). Urbanization and climate change impacts on future urban flooding in Can Tho city, Vietnam. *Hydrology and Earth System Sciences*, 17(1), 379–394.

8. Laukkonen, J., Blanco, P. K., Lenhart, J., Keiner, M., Cavric, B., and Kinuthia-Njenga, C. (2009). Combining climate change adaptation and mitigation measures at the local level. *Habitat International*, 33(3), 287–292.

9. Harlan, S. L., and Ruddell, D. M. (2011). Climate change and health in cities: Impacts of heat and air pollution and potential co-benefits

from mitigation and adaptation. *Current Opinion in Environmental Sustainability*, 3(3), 126–134.

10. Cuny, F. C. (1991). Living with floods: Alternatives for riverine flood mitigation. *Land Use Policy*, 8(4), 331–342.

11. Ward, R. C. (1978). *Floods: A Geographical Perspective*. Macmillan.

12. IFRC (2009). *Climate Change Adaptation Strategies for Local Impacts: Key Messages for UNFCCC Negotiators.* International Federation of Red Cross and Red Crescent Societies.

13. Smith, K., and Ward, R. (1998). *Floods: Physical Processes and Human Impacts.* John Wiley & Sons, Baffins Lane, England.

14. Tingsanchali, T. (1996). *Flood and Human Interaction, Experience, Problems and Solutions.* Professorial Inaugural Lecture. Water Engineering and Management Program. Asian Institute of Technology, Bangkok, Thailand.

15. DeGregorio, M., and Huynh, C. V. (2012). *Living with Floods: A Grassroots Analysis of the Causes and Impacts of Typhoon Mirinae.* ISET-Vietnam, Hanoi.

16. Dozier, E. F., and Yancey, T. N. (1993). *Floodproofing Options for Virginia Homeowners.* US Army Corps of Engineers and Commonwealth of Virginia, Norfolk, Virginia.

17. Wenger, C., Hussey, K., and Pittock, J. (2013). *Living with Floods: Key Lessons from Australia and Abroad.* National Climate Change Adaptation Research Facility, Gold Coast.

Chapter 8

Analysis of Gender Differences in Thermal Sensations in Outdoor Thermal Comfort: A Field Survey in Northern India

Pardeep Kumar and Amit Sharma

Department of Mechanical Engineering,
Deenbandhu Chhotu Ram University of Science and Technology,
Murthal, Sonepat, Haryana, India

kumar.pardeepmech@gmail.com, amitsharma.me@dcrustm.org

Outdoor thermal comfort (OTC) is a vital aspect for healthy livability, efficient working, and society's wellbeing. OTC is negatively impacted by the climate change. Not only do meteorological factors influence people's sensations, but some non-meteorological factors also affect people's senses in outdoor spaces. Among those, gender is one of the most significant factors. This work investigates gender difference in thermal sensations based on questionnaires and objective measurements of meteorological parameters in Haryana's hot semi-arid climate in Northern India. Physiological equivalent temperature index (PET) was employed to determine the effect of meteorological parameters on the

Climate Change and Pragmatic Engineering Mitigation
Edited by Jacqueline A. Stagner and David S.-K. Ting
Copyright © 2022 Jenny Stanford Publishing Pte. Ltd.
ISBN 978-981-4877-97-8 (Hardcover), 978-1-003-25658-8 (eBook)
www.jennystanford.com

human body. The values of PET were calculated using RayMan Pro software. By applying linear regression analysis between the participants' thermal sensations and PET, the neutral temperature range was obtained by selecting $-0.5 \leq$ TSV $\leq + 0.5$ interval on the ASHRAE 7-point sensation scale. The male participants' neutral range was found to be 22.04–32.24°C with a neutral temperature of 27.14°C, while the neutral temperature range for female participants was found to be 25.97–35.77°C with a neutral temperature of 30.87°C. The air temperature was the most significant parameter, followed by solar radiation, impacting both males' and females' thermal sensations. The results show that the female participants feel neutral at a higher temperature than male participants within the thermal environment. The findings of the study could help the various stakeholders associated with the environmental impact assessment and urban designers in developing the outdoor spaces in gender-based educational institutions and other public places such as the shopping mall, parlor, etc. Developing outdoor places would help in mitigating climate change.

8.1 Introduction

Thermal comfort is the state of mind in which a human being feels comfortable in the thermal environment [1]. Due to climate change and global warming, thermal comfort in public places is relatively crucial. Extreme weather events promote people to stay indoors and live a sedentary lifestyle [2]. A sedentary lifestyle harms social, economic, and well-being aspects in society [3]. According to Fong et al.'s study [4], the healthy livability of people is negatively affected by climate change and rapid urbanization. Outdoor thermal comfort (OTC) is essential to promote social and cultural activities. According to Kumar and Sharma's study [5], it was found that improving OTC conditions would attract people to venture outdoor spaces. Consequently, energy use in the buildings would be reduced [6] and help in mitigating climate extremities. OTC is affected by meteorological parameters as well as non-meteorological parameters. According to Middel et al.'s study [7], non-meteorological parameters such as adaptive pattern, thermal preferences, gender difference, seasonal variation, and time of exposure impact the comfort of people. Among all these factors, gender difference is the dominating factor affecting thermal comfort [8]. In previous studies by Jin et al. [8], Middel et al. [7], Lu et al. [9], Makaremi et al. [10], and Oliveira and Andrade [11], it was found that females were more sensitive to the specific outdoor thermal environment than males. To the best of the authors' knowledge, no study has been found in the literature investigating gender difference in the climate zones of India.

The present study analyzes the effect of gender difference in the thermal sensations on outdoor thermal comfort during the winter season in Haryana, which falls in the National Capital Region and Northern India. Based on the motivation, the following objectives have been formulated:

1. To determine the neutral temperature range
2. To determine neutral temperature
3. To check the correlation between the thermal sensations and meteorological parameters

8.2 Methodology

8.2.1 Study Area

This study was investigated at of 28.99°N latitude and 77.01°E longitude in Northern India. According to Köppen-Geiger Climate Classification, this location comes in the hot semi-arid climate (Bsh) zone [12]. The variation of the study area's air temperature and humidity in the past decade is shown in Fig. 8.1. The research methodology framework is shown in Fig. 8.2. The present study was carried out in December 2019.

8.2.2 Data Collection

Data collection was performed by a subjective survey and on-site measurement during five random days in December 2019. A subjective survey was performed using the questionnaire prepared as per the ISO 10551 guidelines [14]. The questionnaire used in the present study is given in Appendix B at the end of the chapter. The questionnaire included personal characteristics such as height, weight, gender, activity, clothing insulation, and age. ASHRAE 7-point sensation scale (Cold (–3), Cool (–2), Slightly cool (–1), Neutral (0), Slightly warm (1), Warm (2), and Hot (3)) was used to record the sensations of the participants.

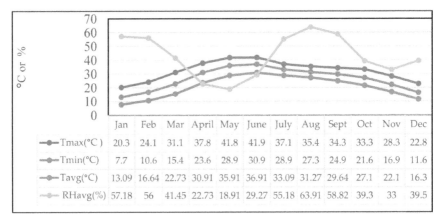

°C or %	Jan	Feb	Mar	April	May	June	July	Aug	Sept	Oct	Nov	Dec
Tmax(°C)	20.3	24.1	31.1	37.8	41.8	41.9	37.1	35.4	34.3	33.3	28.3	22.8
Tmin(°C)	7.7	10.6	15.4	23.6	28.9	30.9	28.9	27.3	24.9	21.6	16.9	11.6
Tavg(°C)	13.09	16.64	22.73	30.91	35.91	36.91	33.09	31.27	29.64	27.1	22.1	16.3
RHavg(%)	57.18	56	41.45	22.73	18.91	29.27	55.18	63.91	58.82	39.3	33	39.5

Figure 8.1 Variation of Air temperature and humidity of the study area from 2010–2019 (*source*: [13]).

Four basic meteorological parameters, air temperature (T_a), relative humidity (RH), wind speed (W_s), and solar radiation (G), were taken into consideration in the present investigation. Air temperature is the "temperature of the air at a point." Relative humidity (RH) refers to "the moisture content (i.e., water vapor) of the atmosphere, expressed as a percentage of the amount of moisture that can be retained by the atmosphere (moisture-holding capacity) at a given temperature and pressure without condensation" [15]. Wind speed is "described as how fast the air is moving past a certain point." [16]. Solar radiation, often called "the solar resource or just sunlight, is a general term for the electromagnetic radiation emitted by the sun" [17]. T_a and RH were recorded using Extech HT30 WBGT meter, while W_s was recorded using Metravi Digital anemometer (AVM-01). All the instruments complied with the ISO 7726 [18]. Solar radiation data were obtained from the meteorological observatory center located at DCRUST, Murthal, Sonepat, Haryana.

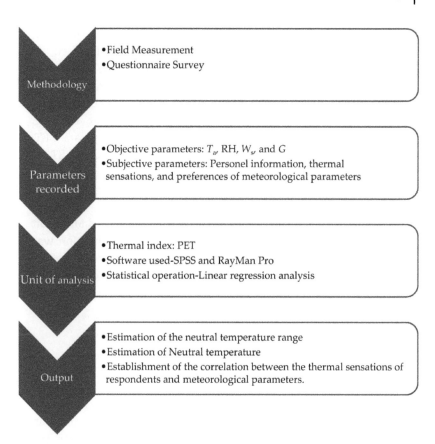

Figure 8.2 Research methodology framework.

8.2.3 Biometeorological Index

In the present study, physiological equivalent temperature (PET) was used to evaluate the participants' thermal comfort conditions in the outdoor environment. PET has been applied most frequently to assess the thermal comfort conditions in the literature [5]. It is based on the Munich energy-balance model for individuals (MEMI), which models the human body's thermal needs physiologically. It is defined as the "physiological equivalent temperature at any given place (outdoors or indoors) and equivalent to the air temperature at which, in a typical indoor setting (without wind and solar radiation), the human body's heat balance is maintained with core and skin temperatures equal to those under the conditions being assessed." This way, PET enables a layperson to compare the integral effects of complex thermal conditions outside with their own experience indoors." [19]. PET was calculated using the RayMan Pro software [20, 21]. RayMan requires the input of meteorological parameters, personal parameters (age, gender, height, clothing, activity), and geographic location information to calculate PET. All the required input was entered into RayMan to calculate PET.

8.2.4 Statistical Analysis

All the collected data was fed into the SPSS software and Microsoft Excel 2016. Linear regression analysis and curve estimation was performed in SPSS software. Spearman correlation was applied to evaluate the preference of meteorological parameters.

8.3 Results and Discussion

8.3.1 Descriptive Characteristics

A transversal survey was conducted to fill the questionnaire (subjective survey). During the subjective poll, 209 questionnaires were supplied by the respondents in an outdoor environment, out of which 185 valid questionnaires were selected for the investigation. In the real world, it is not possible for every participant to give a response to each question in the questionnaire. The questionnaire provided was voluntary in nature. If someone was not interested in filling in all responses in the questionnaire, he/she could leave the response blank. So, the incomplete questionnaire was excluded.

Out of the 185 respondents, 68 were females and 117 were males. Out of the total, 86.4% of the respondents were in the age group of 17–24 years; 11.9% of the respondents found to be in the age group of 25–34 years; 1.1% in 45–54, and 0.5% above 54 years. The majority of respondents were found to be chatting while sitting or standing. The metabolic rate is taken as 70 W. The respondents' values of clothing insulation were taken from ASHRAE Standard 55 (ASHRAE, 2017), corresponding to the respondents' responses. The minimum value of clothing insulation was 1.01 Clo and maximum was 1.3 Clo. The percentage distribution of the TSV by the respondents is shown in Fig. 8.3. The preferences of the meteorological parameters by the respondents are shown in Figs. 8.4 to 8.7. The preferences of the meteorological parameter by respondents were evaluated by applying non-linear spearman correlation test in SPSS software. In the case of male respondents, the preference for solar radiation has the strongest correlation (–0.101) with air temperature, followed by relative humidity (–0.128) and wind speed (0.076). The correlation coefficient sign demonstrated that the higher the air temperature is, the lower the relative humidity and solar radiation.

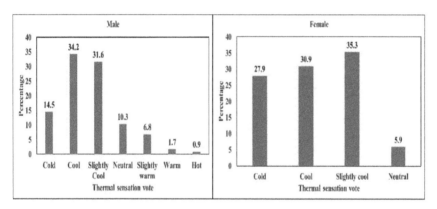

Figure 8.3 Percentage distribution of thermal sensation votes.

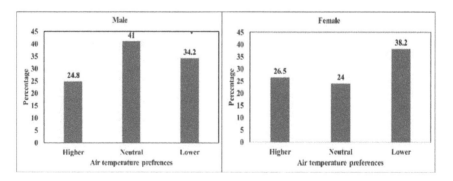

Figure 8.4 Percentage distribution of air temperature preferences votes.

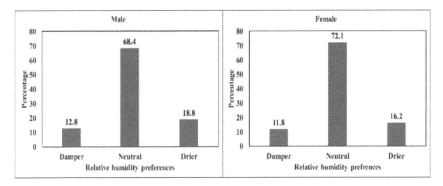

Figure 8.5 Percentage distribution of relative humidity preferences votes.

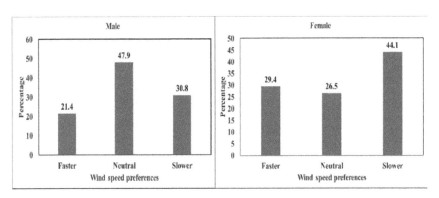

Figure 8.6 Percentage distribution of wind speed preferences votes.

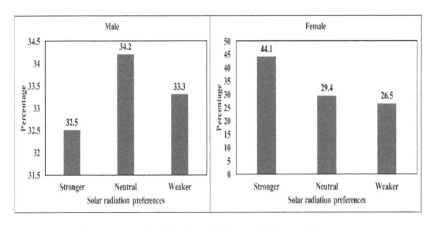

Figure 8.7 Percentage distribution of Solar radiation preferences votes.

While in the case of female respondents, the preferences in solar radiation have the strongest correlation (0.346) with air temperature followed by wind speed (0.204) and relative humidity (−0.076). The sign of correlation coefficient demonstrated that the higher the air temperature is, the lower the relative humidity. The correlation matrix of the meteorological parameters preferences is given in Table A.3 of Appendix A.

8.3.2 Neutral Temperature

It is the temperature at which the occupants feel neutral, that is, occupants are feeling neither hot nor cold while exposed to particular thermal environments [22]. The method developed by de Dear and Brager [23] was employed to find the neutral temperature. By using the developed method, the PET was grouped into 1°C bin. For example, when the MTSV among 88 participants who were exposed to 29–30°C PET is 0.51, it can be regarded as MTSV = 0.51 while PET equals 29.5°C.

8.3.2.1 Male's neutral temperature

While determining male subjects' neutral temperature, a linear relationship was obtained between the mean TSV and mean of PET with $R^2 = 0.54$. A significant correlation was found between MTSV and PET with significance (P-value) = 0.004. Curve estimation between these two can be observed from Fig. 8.8.

Neutral PET was obtained by putting MTSV = 0 in Eq. 8.1.

$$MTSV = 0.098PET - 2.660 (R^2 = 0.54), \text{ significance} = 0.004$$
$$(8.1)$$

The neutral temperature range was obtained by selecting the interval −0.5 ≤MTSV≤ +0.5 in Eq. 8.1. The neutral range was found to be 22.04–32.24°C with a neutral temperature of 27.14°C.

8.3.2.2 Female's neutral temperature

While determining the neutral temperature for female subjects, a linear relationship was obtained between the mean MTSV and the Mean of PET with $R^2 = 0.71$. A significant correlation was found between MTSV and PET with *P* value = 0.001. Curve estimation between these two can be observed from Fig. 8.9.

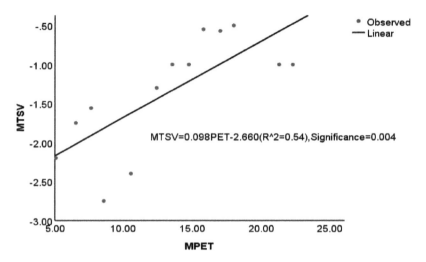

Figure 8.8 Curve estimation between the mean TSV and PET.

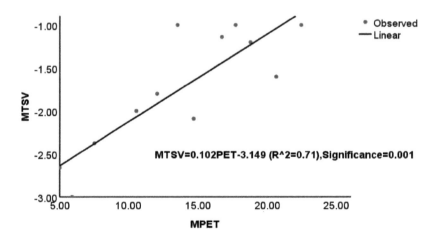

Figure 8.9 Curve estimation between the mean TSV and PET.

Neutral PET was obtained by putting MTSV= 0 in Eq. 8.2.

$$MTSV = 0.102PET - 3.149 \ (R^2 = 0.71), \text{significance} = 0.001$$

$$(8.2)$$

The neutral temperature range was obtained by selecting the interval $-0.5 \leq \text{MTSV} \leq +0.5$ in Eq. 8.2. The neutral range was found to be 25.97–35.77°C with a neutral temperature of 30.87°C. The correlation between the TSV and meteorological parameters are given in Table A.1 of Appendix A. The PET range at various thermal sensations is shown in Table 8.1.

In comparison with previous studies, the neutral temperature for males and females, respectively, was 19.8 and 23.2°C in Harbin, China [8]; 26.1 and 25.2°C in Taiwan [24]; 22.8 and 21.2°C in Wuhan, China [25]; 20.41 and 21.12°C in Melbourne [3]; and 22.15 and 20.79°C for the combined location of Tianjin, China, and West Lafayette, USA [26]. The neutral temperature was determined by combing the environmental conditions, personal details, and perceptions of the people. So the studies carried in various regions have different climate/environmental conditions. The people's cultural and behavioral aspects also play an essential role in the perceptions of people in various regions of the world.

Table 8.1 PET range at various thermal sensations for males and females

Thermal sensations	Assumed TSV	PET range (male)	PET range (female)
Cold	<−2.5	<1.63°C	<6.36°C
Cool	−2.5 to −1.5	1.63 to 11.83°C	6.36 to 16.17°C
Slightly cool	−1.5 to −0.5	11.83 to 22.04°C	16.17 to 25.97°C
Neutral	−0.5 to 0.5	22.04 to 32.24°C	25.97 to 35.77°C
Slightly warm	0.5 to 1.5	32.24 to 42.45°C	35.77 to 45.57°C
Warm	1.5 to 2.5	42.45 to 52.65°C	45.57 to 55.38°C
Hot	>2.5	>52.65°C	>55.38°C

8.3.3 Correlation between the Meteorological Parameters and Thermal Index

In this section, a correlation was established among the meteorological parameters and PET by using Pearson correlation. In terms of male respondents, the strongest correlation was found between the PET and solar radiation ($R^2 = 0.53$). Meanwhile, the strongest correlation was found between PET and wind speed

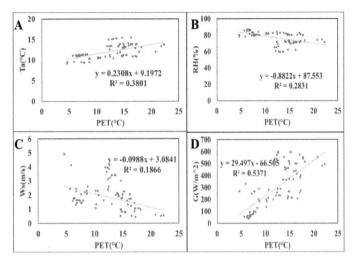

Figure 8.10 Correlation between PET and meteorological parameters for male respondents (A) Correlation between T_a and PET (B) Correlation between RH and PET (C) Correlation between W_s and PET (D) Correlation between G and PET.

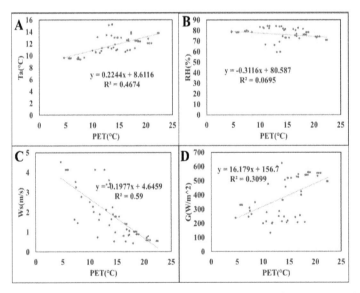

Figure 8.11 Correlation between PET and meteorological parameters for female respondents: (A) Correlation between T_a and PET. (B) Correlation between RH and PET. (C) Correlation between W_s and PET. (D) Correlation between G and PET.

(R^2 = 0.59) in the case of female respondents. The scatterplot of the correlation between PET and meteorological parameters is shown in Figs. 8.10 and 8.11. The correlation matrix generated using the Pearson Correlation is shown in Table A.2 of Appendix A.

8.4 Conclusion

This study investigated gender difference in thermal sensations in OTC in the hot semi-arid climate (Bsh) of Haryana. The thermal index PET was employed to estimate the effect of the thermal environment on human thermal comfort. After taking inference from literature, not only meteorological factors impact people's sensations, but also some non-meteorological factors affect people's sensations in outdoor spaces. Among those, gender is one of the most dominating factors that affect thermal sensations in the outdoor environment. The key findings of the investigated study are as follow:

1. The male participants' neutral range was found to be 22.04–32.24°C with a neutral temperature of 27.14°C.
2. The neutral temperature range for female participants found to be 25.97–35.77°C with a neutral temperature of 30.87°C.
3. The female participants feel neutral at a higher temperature than male participants in the thermal environment.
4. Air temperature, followed by solar radiation, is found to be the most dominant parameter affecting the thermal sensations of both males and females.
5. Solar radiation is found to be the strongest correlation with PET for male respondents, while the strongest correlation of PET was found with air temperature for female respondents.

The findings could help the various stakeholders associated with the environmental impact assessment and urban designers in developing the outdoor spaces in gender-based educational institutions and other public places such as shopping malls and parlors. Developing outdoor places would help people venture out or spend more time outdoors, and buildings, which are closed energy-intensive areas, would be used less, and this would be a step towards mitigating climate change.

Acknowledgments

The National Project Implementation Unit (NPIU) supports this work, a unit of Ministry of Human Resource Development, Government of India, through TEQIP-III Project at Deenbandhu Chhotu Ram University of Science and Technology, Murthal, Haryana, India. The authors thank the meteorological observatory center authorities for providing the global solar radiation data. Special thanks to the respondents who made this survey complete.

Appendices

Appendix A

Table A.1 Spearman Correlation matrix showing the correlation between TSV and meteorological parameters

		Male					Female			
		TSV	T_a	RH	W_s	G	TSV	T_a	RH	W_s	G
TSV	Correlation coefficient	1.000	0.531**	-0.334**	-0.261**	0.313**	1.000	0.474**	-0.333**	-0.331**	0.340**
	Sig(2-tailed)	NA	0.000	0.000	0.005	0.001	NA	0.000	0.006	0.006	0.005
T_a	Correlation coefficient	0.531**	1.000	-0.761**	0.060	0.547**	0.474**	1.000	-0.773**	-0.183	0.582**
	Sig(2-tailed)	0.000	NA	0.000	0.518	0.000	0.000	NA	0.000	0.135	0.000
RH	Correlation coefficient	-0.334**	-0.761**	1.000	-0.176	-0.743**	-0.333**	-0.773**	1.000	-0.062	-0.704**
	Sig(2-tailed)	0.000	0.000	NA	0.057	0.000	0.006	0.000	NA	0.616	0.000
W_s	Correlation coefficient	-0.261**	0.060	-0.176	1.000	0.096	-.331**	-0.183	-0.062	1.000	-0.063
	Sig(2-tailed)	0.005	0.518	0.057	NA	0.306	0.006	0.135	0.616	NA	0.612
G	Correlation coefficient	0.313**	0.547**	-0.743**	0.096	1.000	0.340**	0.582**	-0.704**	-0.063	1.000
	Sig(2-tailed)	0.001	0.000	0.000	0.306	NA	0.005	0.000	0.000	0.612	NA

Table A.2 Pearson Correlation matrix showing the correlation between meteorological parameters and PET

		Male					Female				
		T_a	RH	W_s	G	PET	T_a	RH	W_s	G	PET
T_a	Correlation coefficient	1	-0.867**	0.073	0.594**	0.617**	1	-0.785**	-0.207	0.633**	0.684**
	Sig(2-tailed)	—	0.000	0.433	0.000	0.000	—	0.000	0.000	0.000	0.000
RH	Correlation coefficient	-0.867**	1	-0.333**	-0.669**	-0.532**	-0.785**	1	-0.245*	-0.571**	-0.264*
	Sig(2-tailed)	0.000	—	0.000	0.000	0.000	0.000	—	0.044	0.000	0.030
W_s	Correlation coefficient	0.073	-0.333**	1	0.102	-0.432**	-0.207	-0.245*	1	-0.108	-0.768**
	Sig(2-tailed)	0.433	0.000	—	0.272	0.000	0.090	0.044	—	0.379	0.000
G	Correlation coefficient	0.594**	-0.669**	0.102	1	0.733**	0.633**	-0.571**	-0.108	1	0.557**
	Sig(2-tailed)	0.000	0.000	0.272	—	0.000	0.000	0.000	0.379	—	0.000
PET	Correlation coefficient	0.617**	-0.532**	-0.432**	0.733**	1	0.684**	-0.264*	-0.768**	0.557**	1
	Sig(2-tailed)	0.000	0.000	0.000	0.000	—	0.000	0.030	0.000	0.000	—

Table A.3 Spearman correlation showing preferences of participants regarding meteorological parameters

		Male				Female			
		Preferred T_a	Preferred RH	Preferred W_s	Preferred G	Preferred T_a	Preferred RH	Preferred W_s	Preferred G
T_a	Correlation coefficient	1.000	-0.128	0.076	-0.101	1.000	-0.076	-204*	.346**
	Sig(2-tailed)	—	0.298	0.536	0.414	—	0.418	0.028	0.000
RH	Correlation coefficient	-0.128	1.000	.248*	-0.209	-0.076	1.000	-0.031	-0.131
	Sig(2-tailed)	0.298	—	0.041	0.087	0.418	—	0.739	0.158
W_s	Correlation coefficient	0.076	.248*	1.000	-574**	-204*	-0.031	1.000	-411**
	Sig(2-tailed)	0.536	0.041	—	0.000	0.028	0.739	—	0.000
G	Correlation coefficient	-0.101	-0.209	-574**	1.000	.346**	-0.131	-411**	1.000
	Sig(2-tailed)	0.414	0.087	0.000	—	0.000	0.158	0.000	—

* Correlation is significant at the 0.05 level (2-tailed)
** Correlation is significant at the 0.01 level (2-tailed).

Appendix B

B.1 Questionnaire used in the present study

Outdoor thermal comfort Questionnaire
Note: This questionnaire is completely voluntary.
If you don't wish to complete it or any part of it, you are under no
obligation to do so.

Date   **Time**   **Location**

Gender: Male ☐   Female ☐   **Height**

Age: ☐

Activity: Exercising ☐   Chatting standing ☐   Chatting sitting
Strolling ☐

Any other activity please specify ....

What are you wearing now?

T-Shirt (Short sleeves) ☐   T-Shirt (Long sleeves) ☐   Shorts or
short skirt ☐

Long pants or long Skirt ☐   Sports skirt ☐   Long-sleeve
Shirt ☐   Long-sleeve Sweater ☐

Any other please specify ....

How are you feeling at this moment?

Cold	Cool	Slightly cool	Neutral	Slightly warm	Warm	Hot
−3	−2	−1	0	+1	+2	+3

Thermal preference in regards to microclimate parameters:

Air temperature	Higher	Neutral	lower
Relative Humidity	Damper	Neutral	Drier
Wind speed	Faster	Neutral	Slower
Solar radiation	Stronger	Neutral	Weaker

References

1. ASHRAE, ANSI/ASHRAE Standard 55–2017: Thermal Environmental Conditions for Human Occupancy, ASHRAE Inc. 2017 (2017), 66.

2. F. Salata, I. Golasi, W. Verrusio, E. de Lieto Vollaro, M. Cacciafesta, A. de Lieto Vollaro, On the necessities to analyse the thermohygrometric perception in aged people. A review about indoor thermal comfort, health and energetic aspects and a perspective for future studies, *Sustain. Cities Soc.*, 41 (2018), 469–480. https://doi.org/10.1016/j.scs.2018.06.003.

3. S. Shooshtarian, I. Ridley, The effect of individual and social environments on the users thermal, *Sustain. Cities Soc.*, 26 (2016), 119–133. https://doi.org/10.1016/j.scs.2016.06.005.

4. C. S. Fong, N. Aghamohammadi, L. Ramakreshnan, N. M. Sulaiman, P. Mohammadi, Holistic recommendations for future outdoor thermal comfort assessment in tropical Southeast Asia: A critical appraisal, *Sustain. Cities Soc.*, 46 (2019), 101428. https://doi.org/10.1016/j.scs.2019.101428.

5. P. Kumar, A. Sharma, Study on importance, procedure, and scope of outdoor thermal comfort – A review, *Sustain. Cities Soc.*, 61 (2020), 102297. https://doi.org/10.1016/j.scs.2020.102297.

6. M. Nikolopoulou, N. Baker, K. Steemers, Thermal comfort in outdoor urban spaces: Understanding the Human parameter, *Sol. Energy*, 70 (2001), 227–235. https://doi.org/10.1016/S0038-092X(00)00093-1.

7. A. Middel, N. Selover, B. Hagen, N. Chhetri, Impact of shade on outdoor thermal comfort—a seasonal field study in Tempe, Arizona, *Int. J. Biometeorol.*, 60 (2016), 1849–1861. https://doi.org/10.1007/s00484-016-1172-1175.

8. H. Jin, S. Liu, J. Kang, Gender differences in thermal comfort on pedestrian streets in cold and transitional seasons in severe cold regions in China, *Build. Environ.*, 168 (2020), 106488. https://doi.org/10.1016/j.buildenv.2019.106488.

9. S. Lu, H. Xia, S. Wei, K. Fang, Y. Qi, Analysis of the differences in thermal comfort between locals and tourists and genders in semi-open spaces under natural ventilation on a tropical island, *Energy Build.*, 129 (2016), 264–273. https://doi.org/10.1016/j.enbuild.2016.08.002.

10. N. Makaremi, E. Salleh, M. Z. Jaafar, A. Ghaffarian Hoseini, Thermal comfort conditions of shaded outdoor spaces in hot and humid

climate of Malaysia, *Build. Environ.*, 48 (2012), 7–14. https://doi.org/10.1016/j.buildenv.2011.07.024.

11. S. Oliveira, H. Andrade, An initial assessment of the bioclimatic comfort in an outdoor public space in Lisbon, *Int. J. Biometeorol.*, 52 (2007), 69–84. https://doi.org/10.1007/s00484-007-0100-0.

12. M. C. Peel, B. L. Finlayson, T. A. Mcmahon, Updated world map of the Köppen-Geiger climate classification Updated world map of the Köppen-Geiger climate classification, 2007. www.hydrol-earth-syst-sci-discuss.net/4/439/2007/ (accessed April 11, 2020).

13. https://www.worldweatheronline.com/sonipat-weather/haryana/in.aspx, World Weather Online, WorldWeatherOnline. (2020). https://www.worldweatheronline.com/sonipat-weather/haryana/in.aspx (accessed August 6, 2020).

14. ISO, 10551 Ergonomics of the thermal environment — Assessment of the influence of the thermal environment using subjective judgement scales, (2001). https://www.iso.org/standard/18636.html (accessed November 19, 2019).

15. Relative Humidity: An overview | ScienceDirect Topics, (n.d.). https://www.sciencedirect.com/topics/agricultural-and-biological-sciences/relative-humidity (accessed January 4, 2021).

16. Environmental Monitor|Wind Speed and Direction, (n.d.). https://www.fondriest.com/news/wind-speed-and-direction.htm (accessed January 4, 2021).

17. Solar Radiation Basics Department of Energy, (n.d.). https://www.energy.gov/eere/solar/solar-radiation-basics (accessed January 4, 2021).

18. ISO 7726, ISO 7726:1998 Ergonomics of the thermal environment — Instruments for measuring physical quantities, *ISO Stand.*, 1998 (1998), 1–56. https://doi.org/ISO 7726:1998 (E).

19. P. Höppe, The physiological equivalent temperature: A universal index for the biometeorological assessment of the thermal environment, *Int. J. Biometeorol.*, 43 (1999), 71–75. https://doi.org/10.1007/s004840050118.

20. A. Matzarakis, F. Rutz, H. Mayer, Modelling radiation fluxes in simple and complex environments—application of the RayMan model, *Int. J. Biometeorol.*, 51 (2007), 323–334. https://doi.org/10.1007/s00484-009-0261-0.

21. A. Matzarakis, F. Rutz, H. Mayer, Modelling radiation fluxes in simple and complex environments: Basics of the RayMan model, *Int. J.*

Biometeorol., 54 (2010), 131–139. https://doi.org/10.1007/s00484-009-0261-0.

22. W. Liu, Y. Zhang, Q. Deng, The effects of urban microclimate on outdoor thermal sensation and neutral temperature in hot-summer and cold-winter climate, *Energy Build.*, 128 (2016), 190–197. https://doi.org/10.1016/j.enbuild.2016.06.086.

23. R. J. de Dear, G. S. Brager, Developing an adaptive model of thermal comfort and preference, *ASHRAE Trans.*, 104 (1998), 1–18.

24. C. H. Tung, C. P. Chen, K. T. Tsai, N. Kántor, R. L. Hwang, A. Matzarakis, T. P. Lin, Outdoor thermal comfort characteristics in the hot and humid region from a gender perspective, *Int. J. Biometeorol.*, 58 (2014), 1927–1939. https://doi.org/10.1007/s00484-014-0795-7.

25. J. Huang, C. Zhou, Y. Zhuo, L. Xu, Y. Jiang, Outdoor thermal environments and activities in open space: An experiment study in humid subtropical climates, *Build. Environ.*, 103 (2016), 238–249. https://doi.org/10.1016/j.buildenv.2016.03.029.

26. D. Lai, X. Zhou, Q. Chen, Modelling dynamic thermal sensation of human subjects in outdoor environments, *Energy Build.*, 149 (2017), 16–25. https://doi.org/10.1016/j.enbuild.2017.05.028.

Chapter 9

Urbanization and Food in the Biodigital Age

Secil Afsar, Alberto T. Estévez, and Yomna K. Abdallah

Universitat Internacional de Catalunya,
iBAG—UIC Barcelona (Institute for Biodigital Architecture & Genetics),
Barcelona, Spain

secilafsar@uic.es, estevez@uic.es, yomnaabdallah@uic.es

Unusual problems need unusual solutions. In this sense, mitigating the current global sustainability problems requires sailing into surrealistic creativity as a source of inspiration in invading the intact interdisciplinary fields of research in service of the sustainable style of living as well as the sustainable built environment. In this chapter, the current limitations and drawbacks of the built environment are discussed; the insufficiency of the current style of living and architecture to fulfill human needs of food and energy production Is highlighted. Solutions that are focused on developing self-sufficient, productive home/farm, and edible building materials for temporal sheltering are proposed, including the methodology of implementing these systems and materials, their production technology in the digital age advances, and their reciprocal interrelation with the essential aspects of

Climate Change and Pragmatic Engineering Mitigation
Edited by Jacqueline A. Stagner and David S.-K. Ting
Copyright © 2022 Jenny Stanford Publishing Pte. Ltd.
ISBN 978-981-4877-97-8 (Hardcover), 978-1-003-25658-8 (eBook)
www.jennystanford.com

the design process loop, especially, economy, reproducibility, and sustainability.

9.1 Introduction

As it is known, the Earth is a watery place. Almost 71% percent of its surface is covered with water. This massive water content of the Earth is covering the deeper nested layers of solid matter.

This physical-chemical balance between water and solid content on the Earth's surface is similar to the composition of the human body which contains up to 60% of water. According to H. H. Mitchell (*Journal of Biological Chemistry*, 158), the brain and heart are composed of 73% water, the skin contains 64% water, and even this percentage at the bones is 31%.

The human body is designed to include the water content at constant rates and to sustain the water cycle inside our bodies. So, later it can join the water cycle of the Earth ecosystem as all other creatures do. The eco-system of Earth is composed of self-similar fractal groups. This implies the need for its components to be in a more elementary state. As the holistic scale of Earth's water cycle is attained by the countless living systems, the contribution of all these systems facilitates the regenerative growth cycle. Through the deep understanding of segregation and interdependency of network systems that built up the Earth ecosystem, we need to design our architectural and built environment like it is the part of this ecosystem like how the human body is.

Our current architectural practice produces a built environment that is a "dead load" or nearly lifeless envelopes. In the best case, it can be—ultimately functional—but not sustainable nor ecological. This is due to the absence of "integrity" and the tendency towards "radical objectivity" which leads to a linear design process. Despite the "parametric" attempts to reintroduce the design process as a modifiable "loop," the results are still limited in optimizing built-in aspects of climate responsiveness, user customization (interactivity), and sustainability (eco-friendly). These parametric tools of form-finding, simulation, and optimization have proposed a step to move forward towards multi-faceted study and improvement of the design process.

On the other hand, it still did not reach the "interconnected network of loops" that a design process should be based on.

An integrated loop's network is a hard goal. Our current limited binary tools are not sufficient to uncover all aspects that should be maintained in the urban and architectural design at all once, unless we succeed to reach the multi-dimensional, data processing capacities of our smart artificial minds and to consider all loops inside the design process.

Despite the primitive minds of our ancestors, they were better at balancing intelligently what they "live on" with what they "live from" than modern humans. Their primitive technology and simple tools made from natural sustainable elements enabled the coherent combination of natural agents that were distributed wisely and sufficiently for different functions. When our ancestors managed to discover agriculture, the urban communities were established in their agricultural plots. They lived in it, on it, and with it. The homes were integrated and organized in the agricultural lands. This enabled direct contact with the lands and their crops while maintaining the biocompatibility of architectural building materials, procedures, and performance.

Due to moving from "clusters' sufficiency" of the self-maintained agricultural capsules that include the small plot of agricultural land, the habitat, and the supplementary structures like barns or labs, towards "mass production" and to the capitalized organizations, agriculture gradually began to be considered as an isolated specialized profession, which is separated from everyday life and living-in spaces. The increased population that needed to expand to a wider horizontal area of land caused the diminishing of the self-sufficient agricultural-habitat capsules integration and decrease at the sustainability of the living spaces. This empowered objective functionality that emerged all the plain "less" functionalism architecture has been continued to brutally apply by the post-industrial revolution communities and capitalist ideology. Influenced by the rigidity of machines, specialization, and "distribution" in the post-industrial revolution have resulted not only to insufficient architecture but also in an insufficient lifestyle, which has escalated so far since then.

The concept of the architectural complex with the multi-functions where it is integrated into self-sufficiency activities is should be considered as a norm. A habitat that embraces a human life should include all the functions that guarantee his independence and sufficiency, including at least growing and processing his own food, clothes, remedies, connecting with the surrounding community, and enabling him from monitoring and evolving his cultural activities. Even specialized architectural constructions need this self-sufficiency integrity. For example, agricultural spaces for growing herbs and plants to prepare different chemicals; places for processing, and production of medications; and systems for processing their wastes should be included in hospitals.

Recently, this holistic sustainable integrity of self-sufficient architectural capsules is adopted as the only adequate solution for colonizing Mars, as a future sanctuary from Earth's natural revenge. Designing Mars emergent cities according to the self-sufficient cluster concept will include supporting all human activities through the architecture of this habitat; such as providing climate-responsive and interactive shelters, enabling to grow food and other crops and further processing of these agricultural products and their wastes.

To accomplish the integration of self-sufficiency in urban and architectural design, we need to introduce systems of agriculture as part of our built environment and to overcome the problem of space limitations by clusterization.

9.2   Systems of Food Production Embedded in Architecture (Agricultural Architecture)

Since the industrial revolution followed by exaggeration, technological advancements related to "fast and accessible services and products" for all users "just by a button click" have been effecting the diminishing of customized products and experiences. The "make/repair" concept on everyday life bases was defeated by the readymade commercial solutions that propose easier, faster and no effort solutions. These readymade kits vary in the degree of how much ready they are, starting from not growing veggies and fruits rather preferring to buy them from merchants or throwing instead of repairing or recycling. This tendency can be escalated to reach even humanitarian aspects by creating humanoid robots for fulfilling the social and emotional needs of the modern-futuristic human being.

In order to solve this dilemma, the concept of "less effort and time saving" concepts must be redefined in technology development considering the current state of being lazy and abusive attitude. The "ready-made" term should be replaced by "DIY—do it yourself," buying should be replaced by producing, and capitalism should be replaced by self-sufficiency.

Recently, various trends and socio-architectural movements have emerged to resist the solid concrete dominance on the urban fabric. These socio-environmental activist movements have been calling to embed agriculture and forestry in our living-in spaces either to compensate for the industrial pollution or to cover our shortage of crops due to the decreased rural lands. Planting green-rooftops (Fig. 9.1), hydroponic agricultural methods that need no soil or "land," and embedded mini-greenhouse systems that boost the growth of home-grown crops, are all proposed solutions for combining farming practices with residential architectural spaces.

Figure 9.1 The illustration of a huge urban park with green inter-connected roofs: Alberto T. Estévez, *Green Barcelona Project*, Barcelona, 1995–1998.

Each of these methods depends mainly on the "cluster or pixel" in the global effect resolution, and capitalism holistic centralization cannot be solved in the same centered way but rather by diffusion, collaboration, and miniaturization. The "butterfly wing effect" proved to be the easiest, the fastest, and the most effective method to solve our urban problems.

Moving away from dealing with infrastructures insufficiency for large-scale implementation of productive systems to a portable, and preferring compact and self-sufficient system that is nearly in the size of a small fridge, that can cover the personal consumption of various crops can be one of the solutions.

The ideal to build is self-sufficiency, where each house, each building, each city, tends to be self-sufficient. And therefore, to the self-management of their resources: to get models of buildings that only need the fire of the sun, the wind of the air, the water of the rain (or of the humidity), and the heat of the earth. These would be the genuine "sustaining resources for tomorrow," the four classic elements (Vasel and Ting, 2019), the usual ones. These are the most "natural," sustainable, economical, and democratic resources, and they are renewable and ecological. This is natural intelligence. Later, with the help of artificial intelligence, the housing modules could act almost like companies that have the ability to manage their resources (and wastes) produced by them.

The economy that nature teaches us (bio-learning) has understood natural structures with survival objectives, maximum efficiency, and sustainability after millions of years "testing." While looking for optimal ways of survival in nature, we discovered the organization of cells in different levels of fractality. Of course, there are examples of fractality that can be seen with the naked eye like different plants. Even the sustainability of the planet must also be achieved "fractally," with the sustainability of each country, each region, each city, each house (Estévez, 2020).

The rooftop garden has emerged in the first place as a sanctuary from the dominant feeling of being imprisoned inside a concrete lifeless cube and a way of reducing the drawbacks of CO_2 emissions. Lately, this trend was directed towards producing edible crops to achieve a sort of economic self-sufficiency and growing crops instead of purchasing it. The architectural design of these roof garden applications had to have many parameters to consider including the type of crops and the plant physiology (the length and aggregation form of roots, light intensity, watering rate, and amount), the spatial farmed area, and the seasonal specifications of different crops.

All these aspects controlled the process of implementing the productive rooftops, respecting the structural specifications of the built environment, and manipulating a supply network for plant aggregation. There are numerous successful examples of rooftop gardens that are sufficient on the scale of the micro-community resembled in the one building block habitats (https://desertification.wordpress.com/category/gardening-horticulture/rooftop-gardening/).

The challenges related to this system are the need for collaborative work among the inhabitants and the demand for basic knowledge about forestry and farming. Thus, small-scale projects are emerged to overcome the social interaction associated issues. The cluster scale introduces a clustered greenhouse system that is similar in size to a fridge or an oven (https://www.urbanorganicyield.com/hydroponics-system/), where special ultraviolet light boosts the growth of the crops. Originally these systems were developed for Mars colonization, as an experiment about the ability of plants to grow inside space stations, for instance, a portable Martian greenhouse is currently being developed at NASA (https://eos.org/articles/tests-indicate-which-edible-plants-could-thrive-on-mars). After the COVID-19 crisis, it became obvious that self-sufficiency would always be a must, and the luxury of laziness is not available anymore. In any case, fixing the environmental problems of our own planet first will contribute to be successful while working on creating a life on Mars. These systems highlighted the duality of the production/consumption loop that could be easily performed independently even on a personal scale.

The lack of agricultural knowledge in some communities and the time margins of the productive cycle create the main challenge for these systems. Despite these challenges, these agro-architectural systems are gaining social appeal currently, thanks to the eco-friendly and sustainable cultural concepts that boost the interaction and integration between humans and nature.

Achieving self-sufficiency on the cluster scale will emerge into a redefined urbanization in principles and practices. The introduced concepts of self-reliance and autonomous production as part of architectural spaces will boost the overall urban green fabric and limit the climatic disruption that our planet is facing lately.

9.3 Future Scenarios: 3D Printing for Edible Architecture between Urgent and Emergent

As mentioned previously in this chapter, the post-industrial revolution witnessed the rise in the mass-production with the product assembly lines. Despite the increase in the global population, consumption-oriented growth strategies of companies have triggered to focus more on the ease of production and economic viability of the process than the interaction between users and products. This tendency was causing a shift towards more mechanistically processes and replicable and replaceable product generation.

Among ideology makers, the dominance of mechanization in architectural design, was not welcomed by the traditional craftsmen, starting from the "Art and Crafts Movement," led by John Ruskin, as a "rebellion" against cheap mass production, new design approaches have started to develop based on exertion and craftsmanship. In contrast to mass production units, this labor-based approach was inspired by nature at least in form, materials, and production processes.

Consequently, in the early 1890s, the "Art Nouveau" reintroduced the integration of the most influential forces in the world: art and nature. Art Nouveau's significance and effect extended far beyond any prior attempts to integrate art and nature. Perhaps, due to the sharp and mechanical reflection of industrialization towards the environment; the soft details, flows, and natural elements into the creation process started to be more effective in design and architecture taking a further step towards design efficiency that is informed by natural laws (Orman, 2013).

Gaudí, as the proto-surrealist architect, in the context of Art Nouveau created dream-like buildings such as Casa Milà, Casa Batlló, Park Güell, and the Sagrada Familia temple. His "plasticized" stone designs give the feeling of flowing softly and freely, defeating their material limitations (Fig. 9.2).

This tendency was followed by one of his biggest admirers, Salvador Dalí who was searching for a break from "logic and rationalism." "The architecture of the future will be soft and hairy.

It is already soft with Gaudí and it will be hairy because of Dalí, and then I will also look at the creative power of Gaudí" (Dalí, 2000).

Figure 9.2 From left to right, Antoni Gaudí, Casa Milà (La Pedrera), Barcelona, 1906–1911); Antoni Gaudí, Park Güell, Barcelona, 1900–1914 (photos: Alberto T. Estévez).

Despite Dali's irrational dream, he introduced the concept of performance and sufficiency of architecture, as an edible shelter. He simulated the early life human shelter, the womb, the soft adjustable capsule that shelters which can deliver oxygen, food, and sufficiently ejects the wastes: "Not feeling cold or heat, no thirst, hunger nor pain. When the human being in the womb has everything: uniform temperature, food, without loud noises, nor knocks or tears, because everything is cushioned (...), wrapped up in the womb, (...) the vitelline skin that surrounds us (...) it is perfect, comfortable and grows to measure (...) Dalí himself also spoke about his intrauterine memories" (Estévez, 2005).

Thus, at the beginning of the 20th century, Gaudí introduced forms, and years later Dalí understood and explained it as edible architecture (Dalí, 1933). In mid-century, when he prophesied in Park Güell, genetics would make it happen (Dalí, 1956). Then, at the beginning of the 21st century, we began to scientifically try to work on what could be the genetic architecture: "end up creating a complete living house. A tree with heating. (...) Buildings whose walls and ceilings grow with their own flesh and skin (Fig. 9.3), or at least with plant textures, which genetics is able to develop, including shining heating coming through the veins delivering the oxygen necessary for breathing" (Estévez, 2003).

Figure 9.3 Left: Ludwig Mies van der Rohe, *German Barcelona Pavilion*, Barcelona, 1929, one of the most classic examples of modernist rational-functionalist architecture. Right: Alberto T. Estévez (with Marina Serer), *Genetic Barcelona Pavilion ("Ceci n'est pas un pavillon")*, Barcelona, 2007, soft, edible, genetic manifesto project remodeling of the Mies van der Rohe Pavilion in Barcelona (photos: Alberto T. Estévez).

Nowadays, while fighting against the irresponsible usage of natural resources, many researchers have been developing alternative systems in order to use natural resources efficiently and decrease global emissions. The search for more than informed forms by nature and optimized structures are adding the performance, and productivity of architectural spaces to the urban context.

9.3.1 Edible Architecture for Temporal Uses

Through the regenerative architecture aspect, buildings are in constant conversation with the natural world (Armstrong, 2009). Following the Earth's model of the renewable infinite life cycle where nutrients are endlessly recycled through different interconnected loops of different reciprocal relations, for example, the waste outputs of one organism become nutrient inputs of other organisms. Following the same performance pattern, architecture should be conceived as a renewable regenerative organism, not permanent in its formal, physical, and chemical state and not temporal in its existence. But ever-changing, morphing and evolving in accordance with its interaction with the micro and macro environment, this concept redefines the architectural design process, and consequently the entire urban fabric, with the condition of capacity for polymorphous adjust-ability and commitment to synchronization accuracy and safety criteria. This implies dealing with the architectural unit as a living creature that grows, decays and reintroduces to life in another form. One way of realizing this edible living shelter is developing its materiality to join the life cycle of the eco-system. Agro polymers which are the animal and plant-based renewable raw biomass can degrade to simple substances by other organisms.

Degradation and digestion are similar in terms of the fact that both break down complex organic substances into simple compounds (Eqs. (9.1)–(9.3)). Therefore, inside the edible architecture context, these agro polymers can be developed as part of temporary architectural elements or edibles. So, they can leave organic by-products behind as a part of the natural cycle when they decay or being digested. This perspective leads to a new way of making and using resources by preventing the one-way energy transfer, from the ecosystem to human-made environments, rather convert it to a reversible process. In (Fig. 9.4), the authors exhibit an experimentation on developing an agro polymer as an architectural building material that supports the growth of mycelium.

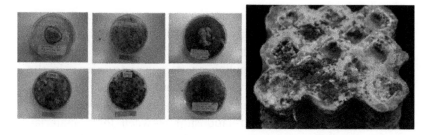

Figure 9.4 From left to right, mycelium growing experiments with various lignocellulosic biomass and chitosan; 3D printed clay for mycelium growing, Secil Afsar and Doruk Yildirim, by using FabLab, Barcelona 3D print pasta extruder, April 2020.

Anaerobic Biodegradation Equation

$$C_{polymer} \longrightarrow C_{residue} + C_{biomass} + CO_2 + H_2O \tag{9.1}$$

Aerobic Biodegradation Equation

$$C_{polymer} + O_2 \longrightarrow C_{residue} + C_{biomass} + CO_2 + H_2O \tag{9.2}$$

Enzymatic Hydrolysis Equation

$$n(C_6H_{10}O_5) + nH_2O \rightarrow n(C_6H_{12}O_6) \qquad (9.3)$$

Equations (9.1), (9.2), (9.3). Equations of biodegradation and hydrolysis process indicating that degradation and digestion are similar in terms of the fact that both break down complex organic substances into simple compounds.

On the other hand, the lack of mechanical properties alongside insufficient environmental stability including environmental interaction is limiting the direct use of agro polymers in architectural building applications. However, their performance can be increased by developing biocomposites and fabricating them in suitable geometries. So, the ephemeral structures made by using these agro polymers can take place as a part of a new architectural approach by their sustainability and they can join the nutrition, hence the energy cycle inside nature.

9.3.2   3D Printing for Edible Architecture

The mechanical properties of these composites made from agro polymers also strongly depend on small changes in its molecular organization, which can be governed by the fabrication methods, rather than depending on its chemistry alone. In the rheological aspect, these soft materials deform or flow in response to applied forces or stresses. To grasp more comprehensively this complex system, the relationship between form, material, and structure should be computed analogically and digitally.

The agro polymers, as part of the structural elements of animals or plants, are generated during their growing processes. During the growth, the formation process leads to the generation of materials hierarchically due to the adaptation of each microstructure to its local needs. This hierarchical adaptation can be affected by not only the internal conditions but also external conditions like temperature, moisture, or external forces.

This bottom-up approach can be supported by 3D printing fabrication techniques. Agro polymers can be fabricated in various geometries and shapes (Fig. 9.5). The integration between material, form, and structure, proposes an alternative strategy to the conventional form-creating process.

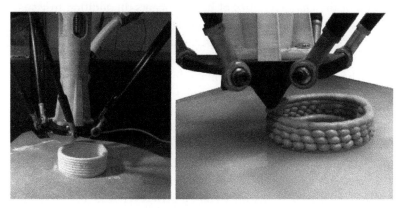

Figure 9.5 From left to right, 3D printing process of chitosan, cellulose microfiber, casein biocomposite, project of Secil Afsar by using FabLab, Barcelona 3D print paste extruder, March 2020; 3D printing process of alginate, eggshell biocomposite, the recipe of Fabricademy, Barcelona, 3D printing process by Secil Afsar with FabLab, Barcelona 3D print paste extruder, March 2020.

However, due to the non-existent material data sheets and inadequate and complicated application processes, the direct usage of agro polymers has been limited. The data about these "edible architectural materials" in their sufficient geometric forms and chemical composition should be optimized. In this regard, the 3D printing paste extrusion is one of the promising techniques to generate sufficient geometric forms of agro polymers in their composite form.

Moreover, with the help of integrating 3D printing paste extrusion techniques, the generative design tools can also join this paradigm which leads to developing customizable edible architectural products and converting agro polymers as a strong competitor against mass-produced, conventional construction materials.

9.3.2.1 Food and architecture mutability

The duality in between hard and soft, interior, and exterior, buildable and unbuildable terms describe the relations at food and architecture's mutability (Horwitz amd Paulette, 2004). Texture, proportion, hierarchy, balance, strength, color, and harmony are some of the characteristics of both concepts. In addition to visual and tactile sensory experiences, designing the texture of the food can develop a new sensory experience and allow us to sense the form while tasting, chewing, or swallowing.

The main parameters to make foods more attractive can be: taste, smell, texture, structure, and composition. Compared to conventional food production techniques, 3D print technology introduces a brand-new approach by giving design opportunities from microscopic to macroscopic level and developing not only structure but also texture and the taste of the foods and introducing them as part of the built environment or as an edible building material. By playing the micro and macro level design parameters, new sensations for eating can also be created.

One example is Barcelona-based company Novameat, which is taking advantage of this opportunity to create a 3D printed plant-based steak (Fig. 9.6). Although nowadays, there are plenty of companies that have been working on plant-based alternatives for meat in the market, Novameat is using an FDM 3D printer to generate detailed layer design and to mimic the texture and appearance of whole beef muscle cuts. Thus, by using the micro-extrusion technique, rather than giving the "bulky" feeling of meatballs, this plant-based alternative is offering the sensation of eating a steak to customers.

Figure 9.6 Left, 3D printed plant-based steak; right 3D printed pork, NovaMeat, Barcelona.

Besides, the Netherlands-based company Upprinting Food and Barcelona-based project called Look Ma, No Hands are offering healthy food alternatives by giving the value back the discarded foods. Therefore, by redesigning foods, which have still nutrition values and high value of fiber contents, they are contributing to the sustainable food cycle and redefining the shape of the food.

In this regard, generative design tools can carry this approach one step further. They can give a chance to develop more complex geometries. Using the parametric design tools, allow us to design each layer specifically. These programs give users the chance of generating customized slicing parameters independently from any predetermined program interfaces.

The other noteworthy advantage of these parametric platforms is their language that can be directly transferred to the robotic arm control (Fig. 9.7). Currently, one of the popular approaches in digital design and fabrication technologies is to use a 6-axis robotic arm for 3D printing. In the future, this 3D print with 6-axis robotic arm technology can be adapted to food production and more complex shapes can be printed faster.

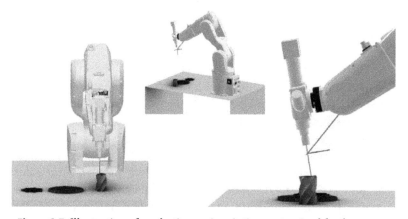

Figure 9.7 Illustration of a robotic arm is printing customized food.

There are still many possibilities to be developed in order to use the advantages of 3D printing food technology, comprehensively. In this regard, not only 3D printing paste extrusion but also other printing techniques, like powder-based printing technologies, should be joined to this game. Later, these technologies can be fed from various disciplines, such as geometrical solutions used in architecture or finite element analysis used in material science.

So, it can not only help us to find solutions for environmental, social, and economic problems in the food supply systems under the discourse of sustainability but also give the opportunity to choose the shape and texture of the food like we choose the taste and nutrition values.

9.3.2.2 Scaling-up and moving beyond

Inside the context of edible architecture, nowadays, there are plenty of projects that have been scaling up for the applications of the urban fabric. The materials which can join the nutrition, and therefore the energy cycle, have begun to take place in temporary architectural projects at different scales and various places.

The structures made from these materials might be used to create the whole or some parts of the structure. Currently, with the help of generative design tools the recursive models, fractals, and coral-like shapes can be embedded in architecture. This time the soft details and flows of the forms like in Art Nouveau are meeting with natural materials. Consequently, Dalí's prophecy about the term "soft and hairy," is beginning to be seen in contemporary architecture and design.

The two examples which have the potential to allow an adaptation of edible architecture can be directly given. The world's first 3D printed neighborhood project for people at risk like homeless people has been building in Tabasco, Mexico Through 3D printing robotic technology, curved walls and lattice structures were printed to increase the airflow. With the help of this technology, single house can be printed with 24 h of print time around several days. Therefore, this technology generates a chance to not only print in higher quality and economically favorable than conventional constructing techniques but also achieving faster results.

Although in this project cement is used as a material for creating long-term housing opportunities, the same aspect can be used to develop temporary structures for disaster zones or on the streets to use as a shelter for homeless people during the winter.

The other example is the mission to set humans' foot on Martian soil. The most popular claim to create a life at Mars is to send construction robots for 3D printing which will prepare the base for human needs before humans are landed. According to Elon Musk, who is the CEO of the company SpaceX, "machines to produce fertilizer, methane, and oxygen from Mars' atmospheric nitrogen and carbon dioxide and the planet's subsurface water ice, and materials to build transparent domes for crop growth" need to be considered in advance.

There are already many projects that have been generating considering possible future habitats such as the NASA's International 3D Printed Habitat Challenge or The Index Awards. The project called Hassell, one of ten short-listed groups' projects in NASA's International 3D Printed Habitat Challenge in 2018, proposed to use material named regolith which can be found at Mars surface and to bind it with heat.

Due to the high weight, carrying concrete from the Earth causes issues (Savage, 2017). Thus, the alternative technique is to use the material which is already there. "All we have to do is develop the technology to use that aggregate that already exists and somehow bind that aggregate together," said Robert P. Muller, who is a senior technologist at Swamp Works, which is cofounded at NASA's Kennedy Space Center. In light of this, edible architecture materials can be a promising solution to develop composite from regolith with their lightweight, printability, and reinforcement of the nutrition cycle.

The other project, named Marsha, Index Award 2019 Winners' project suggested the egg-like shape structure and developed a composite from regolith and PLA. The common points of these prospective projects can be turning the existing material at Mars to suitable for three-dimensional printing and finding the most adequate design. Because of the thin atmosphere, all the air inside of the habitat wants to expand outwards. Thus, the most structurally optimized forms at these projects were generally designed considering enhancing horizontally through outwards.

9.3.3 Design as an Output vs. Design as a System

The term "Anthropocene" defines an era in which humans have an accelerated and intense effect on the world's ecosystem. Especially after the late nineteenth century, humans have started to realize that this gradually increasing consumption and the irresponsible usage of natural resources cause a great degree of environmental damage on Earth (Zalasiewicz et al., 2011).

Before these changes become irreversible, many new approaches have been emerging to find solutions to the ongoing ecological crisis. In the late 20th century, the eco-design aspects were brought out to companies as a new trend.

Later, this environment-focused aspect became insufficient due to not giving enough consideration to economic and social outcomes. Although these "go green" tendencies are very valuable on the paper, in order to make them practically applicable in the long-term, the social and economic value generation should be also included inside the process design (Diehl and Crul, 2008).

In this regard, the Design for Sustainability (D4S) aspect claims to be more applicable since it already includes these elements inside of its definition (Diehl and Crul, 2008). Moreover, over a decade, the popularity of the circular economy has been increasing inside the discourse of sustainable development. Although the circular economy is an economic system, due to its sustainability goal, it considers the design of the whole system and creates a road map starting from the idea generation of the product or service to the attitude of the end-users (Fig. 9.8).

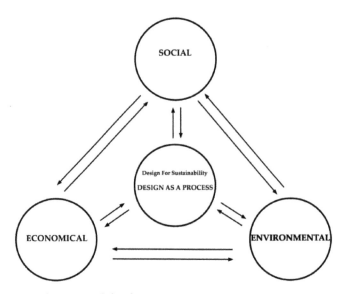

Figure 9.8 The sustainability loop.

Nevertheless, this multifaceted and dynamic system still has difficulties in accomplishing its "closing the loop" goal. One of the reasons for this can be the complexity of the system itself. In contrast to the linear economic model, this complex system requires strategic design by considering inputs and outputs of all process steps (Sarasvathy, 2001).

The second reason can be the current over-consumption tendency of society versus limited resources on Earth. Thus, from the beginning of the project generation process, the availability of resources, shelf life, decomposition time needs to be investigated well.

Considering all these, not only governmental institutions need to support the sustainability aspect, but also engineers, architects, and designers need to understand the whole system to prevent the possible negative environmental, social, and economic outcomes. The iterative steps for process design with feedback loop systems should be constructed. Since the design thinking is a nonlinear process, the data exchange from all the branches helps to strengthen the system. So, the knowledge and experiential intelligence transfer of engineers, architects, and product designers with each other and also with the policy-makers can help to understand the problems and challenges from the roots.

9.3.3.1 An inquiry of architects

Timothy Bloxam Morton, who is a professor at Rice University and a member of the object-oriented philosophy movement, uses the term "hyper-objects" to define the concepts which have massive scale and long-term effects, such as climate change or biosphere. Due to the fact that humans are having difficulties to relate "hyper-objects" that exceed the perceptive capacities of humans and having a tendency to downgraded non-human things to human-scale, inevitably, they are struggling to comprehend these long-term effects and massive scales (Morton, 2016). So, they cannot imagine the CO_2 that they breathe can be produced in a faraway petroleum refinery 20 years ago.

Architects have been facing a hard time in mainstreaming conversations of ecology and humans. However, after the increased effects at the climate crisis and integration of biology to design, they also have begun to constantly ask themselves about how to establish a sustainable environment for human living spaces, which do not separate the human ergonomy from the natural cycles.

An American architectural and design theorist Benjamin H. Bratton suggests that, from now on, half of the urbanists and architects need to transfer their knowledge and experiential intelligence into the design and programming of new software, program, or language to develop the systems and structures. So, the other half can continue to do traditional architecture, while this half can do content management (Bratton, 2019).

This content management developed by architects can be very valuable since they have been working on understanding an individual's experiences with their environments for a long time. So, architects and urbanists need to put their chairs next to the table where they can speak to contribute the definition of a sustainable system and to build a bridge between individuals and the environment.

9.3.3.2 Foodshed and the strategic urbanism

Global food production creates one of the greatest threats to the ecosystem. While food is crucial for our lives, if its system is not developed well, it gives a great impact on our environmental problems. The food supply system, from growing to processing, packing, and transportation is creating greenhouse gases which are approximately responsible for 26% of global emissions. In contrast to this, 30% of the food produced globally is wasted every year (Hanson et al., 2018).

The food supply system is a network of technologies, processes, and people. If this multifaceted system is established in the same geographic location and from growing to the composting process the food travels inside this region, this area is named "foodshed." Developing foodshed and reducing food miles helps to decrease the total green gas emission that warms our planet. Besides, while people can reach fresher food, they can also support local producers and help to conserve regional biodiversity (Karp and Kerrick, 2015).

The term "foodshed" was first used by Walter P. Hedden in his book *How Great Cities Are Fed* in 1929. The book was inspired by the sudden thread during the worldwide strike in the early 20th century. He was pointing out that the public agencies of New York became conscious of how they are defenseless at the distribution of the food and weak against the complexity of the food transportation system.

Although the technology, politics, and eating habits have been transforming, the main argument of how our cities are fed today are remarkably related to concern in the early 20th century. Continuing from the example of New York, after 9/11 the entrance of the trucks which deliver the foods to the center was blocked for several days. This crisis once again showed that, especially for big cities, the city's food supply system was limited if their global or national food supply system was cut off.

Not only examples in New York City in the past but also the current outbreak of the global pandemic, the problem of the food supply system for cities popped up again. Therefore, while constructing a city plan, architects and urbanists can separate special areas for growing food as a part of the buildings or small farming areas that can be located close to the city center.

Then the eating from the "foodshed," can help to keep farms viable and in business. The wildlife habitats can be protected which develops a healthy ecosystem inside the cities. Local agriculture will diversify a greater variety of crops and animals. This will sustain biological stability and enhance diversity. Last but not least, the decrease in CO_2 emission related to a long transportation chain will be observed, and cities will be centralized by the less need for invisible entities across the world.

9.3.3.3 3D print for sustainable urban food cycle

Like every new technology, initially, it may be hard to comprehend the role of 3D printing food inside the sustainable food production system. "When people first heard about microwaves, they didn't understand the technology," said Lynette Kucsama, Chief Marketing Officer at Natural Machines.

Under the title of the environmental contribution of this technology, several benefits can be directly counted. For instance, by 3D printing, the problem of wasting food can be reduced due to only the required amount of raw materials being used to make food. In addition, without the need of any chemical additives, by vacuuming or cold-storing the capsules, food can be stored for a long time. Later by taking advantage of printing with special textures or simply adding some spices, food still can be tasty. Last but not least, since this technology can be easily accessible from local places or many people can print their food from home, lesser transportation will be needed.

Furthermore, not only less transportation will be needed but also many people who are struggling to reach food will easily access healthy food. Also, with the help of predetermined recipes and customized production techniques, this new technology holds great promise for future healthy generations. There are great deals of people who have been fighting against diseases like diabetes or cholesterol. By playing with the taste perception of food with distributing the sweetness or salt heterogeneously in specific places, less amount of salt or sugar can be used to give the same taste.

The other benefit with respect to health is, this technology can give information about the exact composition of the food. Athletes, pregnant women, or people who prefer special diets, such as gluten-free, vegan, or non-allergic foods, can also track how much calories, nutrients, or vitamins they are taking. Moreover, elderly people or patients with swallowing issues can also take advantage of this customized food technology.

Beyond environmental and social gain, this technology can create a positive economic impact by closing the gap between small and large businesses and bringing down the cost of food production. Besides, many economic resources required to solve food supply system problems will be preserved. Lastly, this technology can boost new business opportunities in the food industry and create a new paradigm against conventional food production.

All in all, considering the environmental, social, and economical benefits, 3D print food has a significant role in developing the circular economy inside the discourse of sustainable development and offers a promising way of help for solving the food supply system problems.

9.3.4   Biomanufacturing the Future

Ultimately, in short, biomanufacturing is a type of manufacturing or biotechnology that uses biological systems to produce biomaterials and biomolecules for use in the medical and food industry. Biomanufactured products are recovered from natural sources, such as animal or plant cells. The cells used during the production may have been naturally occurred or derived using genetic techniques. Since the year 2000, we began to apply this with architectural objectives, founding the Genetic Architectures Research Group & Office, and the Master Degree of Biodigital Architecture, at the ESARQ, the School of Architecture of UIC Barcelona (Universitat Internacional de Catalunya). Research, practice, and teaching come together on architecture and design, with the application of biology and digital tools, where natural intelligence, artificial intelligence, bio-learning, machine-learning, bio-manufacturing, digital-manufacturing, and digital organicism are keywords.

We are exploring knowledge frontiers applied to architecture, design, art, and engineering, through interdisciplinary endeavors with bio and digital techniques.

Urbanization and food in the biodigital age, in our age, today! Only a few things have so much potential to solve the problems of sustainability as biomanufacturing. That is when "biomanufacturing the future" becomes a necessity.

Definitions

- **Agro polymers** are the animal and plant-based renewable raw biomass.
- **Biocomposite** is a material formed by a matrix and a reinforcement of natural fibers.
- **Biodegradation** is the process of breaking down complex compounds into simpler parts naturally.
- **Degradation** is the process of breaking down complex compounds into simpler parts naturally or artificially.
- **Decomposition** is the process of breaking down organic compounds into simple parts naturally.

- **Digestion is** absorbing the valuable substance for the body, using chemicals (enzyme) or physical (teeth) techniques to break foods.
- **Micro-extrusion** is an extrusion process that material pushed with the submillimeter range of nozzle size.
- **Printability** is related to the viscosity of the material which should be low enough to extrude and high enough to preserve its shape after the extrusion.

References

Armstrong, R. Architecture that repairs itself? [YouTube Video] TedTalks, 2009. https://www.youtube.com/watch?v=nAMrtHC2Ev0&t=28s.

Baldassarre, B., Calabretta, G., Bocken, N., Diehl, J. C., Keskin, D. The evolution of the strategic role of designers for sustainable development. Conference: Academy for Design Innovation Management Conference, London, Jun. 2019.

Bratton, H. B. *Iphone City*. John Wiley & Sons Ltd., 2019.

Bocken, N., Morales, L. S., Lehner, M. Sufficiency business strategies in the food industry–the case of oatly. *Sustainability*, Jan. 2020.

Clark, G., Kosoris, J., Hong, L., Crul, M. Design for sustainability: current trends in sustainable product design and development. *Sustainability*, Sept. 2009, pp. 410–423.

Dalí, S. Posición moral del surrealismo. *Hèlix*, 10, 1930.

Dalí, S., De la beauté terrifiante et comestible de l'architecture Modern Style. *Minotaure*, 3–4, Paris, 1933.

Dalí, S., Manifesto-Conference, Park Güell, Barcelona, 29 Sept., 1956 (Manuscript preserved in the Fundació Gala-Salvador Dalí Collection, Figueres).

Dalí, S., Los cornudos del viejo arte moderno. *Tusquets, Barcelona*, 2000 1956.

Diehl, J. C., Crul, M. Design for sustainability (D4S): Manual and tools for Developing Countries. *7th Annual ASEE Global Colloquium on Engineering Education*, Cape Town, Oct. 2008.

Estévez, A. T., Genetic architectures. In Estévez, A. T., et al., *Genetic Architectures*, SITES Books/ESARQ (UIC), Santa Fe/Barcelona, 2003, p. 17.

Estévez, A. T., Biomorphic architecture. In Estévez, A. T., et al., *Genetic Architectures II: Digital Tools and Organic Forms*, SITES Books/ESARQ-UIC, Santa Fe/Barcelona, 2005, p. 62.

Estévez, A. T., Sustainable living? biodigital future!. In Stagner, J. A., Ting, D. S.-K. (eds.), *Sustaining Resources for Tomorrow*, Springer, Berlin, 2020.

Godoi, F. C., Bhandari, B. R., Prakash S., Zhang M., *Fundamentals of 3D Food Printing and Applications.* Academic Press, 2019, pp. 1–18.

Hanson, C., Lipinski, B., Friedrich, J., O'Connor C., James K. What's food loss and waste got to do with climate change? A Lot, Actually. World Resources Institute, 26 Sept. 2018. www.wri.org/blog/2015/12/whats-food-loss-and-waste-got-do-climate-change-lot-actually.

Horwitz, J., Paulette S. Eating architecture. May, 2004.

Jawaid, M., Sapuan S. M., Alothman, O. Y. *Green Biocomposites Manufacturing Properties. Green Energy and Technology*, Springer International Publishing AG, 2017.

Steffen, A. D. World's First 3D Printed Neighborhood Is Absolutely Gorgeous. 18 Feb. 2020. https://www.intelligentliving.co/worlds-first-3d-printed-neighborhood-is-absolutely-gorgeous/.

Kaplan, S. *Praise for The Business Model Innovation Factory: How to Stay Relevant When The World Is Changing.* John Wiley & Sons, 1st ed., 27 Apr. 2012, pp. 17–32.

Karp, K., Kerrick, B. How Great Cities Are Fed. Heritageradionetwork, 19 Jan. 2015.

Morton, T. Dark Ecology: For a logic of future coexistence. 2016.

Noort, M. W. J., Bommel, K., Renzetti, S. 3D-Printed Cereal Foods. Cereal Foods World, Nov. 2017.

Norman, D. A., Stappers P. J., DesignX: Complex sociotechnical systems. *She Ji J. Des. Econ. Innovation*, 2015, pp. 83–106.

Orman, B. Art Nouveau & Gaudí: The Way of Nature. *JCCC Honors J.*, 4(1), Article 2, 2013. http://scholarspace.jccc.edu/honors_journal/vol4/iss1/2.

Ritchie, H. Food production is responsible for one-quarter of the world's greenhouse gas emissions. Our World in Data, 6 Nov. 2019, http://ourworldindata.org/food-ghg-emissions.

Rizos, V., Behrens, A., Kafyeke, T., Hirschnitz-Garbers, M., Ioannou, A. The circular economy: Barriers and opportunities for SMEs. CEPS Working Documents, Sept. 2015.

Sarasvathy, S. Causation and effectuation: Toward a theoretical shift from economic inevitability to entrepreneurial contingency. The Academy of Management Review, Apr. 2001.

Savage, N. To build settlements on Mars, we'll need materials chemistry. *Chem. Eng. N.*, 96(1). 27 Dec. 2017. https://cen.acs.org/articles/96/i1/build-settlements-Mars-ll-need.html.

Vasel, A., Ting, D. (eds.). *Air, Water, Food, and Energy—The Four Life-Supporting Elements*. Taylor & Francis/CRC, London, 2019.

Zalasiewicz, J., Haywood, A., Ellis, M. The anthropocene: A new epoch of geological time? *Philos. Trans. R. Soc.*, 2011, pp. 835–840.

https://desertification.wordpress.com/category/gardening-horticulture/rooftop-gardening/.

https://www.urbanorganicyield.com/hydroponics-system/.

https://eos.org/articles/tests-indicate-which-edible-plants-could-thrive-on-mars.

https://www.hassellstudio.com/project/nasa-3d-printed-habitat-challenge.

Chapter 10

How Hydrogen Can Become a Low-Risk Solution for a Climate-Neutral Denmark by 2050

Katarzyna Agnieszka Wierciszewska and George Xydis

Department of Business Development and Technology,
Aarhus University, Herning, Denmark

gxydis@btech.au.dk

This chapter discusses carbon dioxide emission in Denmark and its impact on the realization of the sustainability targets. It analyses the automotive industry in order to determine whether hydrogen cars could have an impact on Denmark's achievement of sustainability targets. However, as hydrogen technology has not been yet deployed satisfactorily, the increase in the number of electric vehicles (EVs) in Denmark was analyzed to determine whether there was any link between the number of charging stations and the number of EVs over the recent years. This has shown that the more charging stations there were, the more EV cars were registered. Based on this assumption, it was concluded that there is a possibility for an increase in the number of hydrogen cars, when there will be an increase in the hydrogen fueling stations. Additionally, the predicted number of hydrogen

Climate Change and Pragmatic Engineering Mitigation
Edited by Jacqueline A. Stagner and David S.-K. Ting
Copyright © 2022 Jenny Stanford Publishing Pte. Ltd.
ISBN 978-981-4877-97-8 (Hardcover), 978-1-003-25658-8 (eBook)
www.jennystanford.com

cars by 2030 was used to calculate the expected reduction of CO_2 emissions. Lastly, the discussion revealed that there is a need for a larger increase in the number of hydrogen cars in order to meet the sustainable targets, based on the national energy and climate plan.

10.1 Introduction

Both engineering and economics confront the environment as a complex asset that offers various services. Although this is a very special asset, the environment, above all, ensures our very existence. As with every other asset, humans want to prevent the undue depreciation in order to ensure that environment will continue to satisfy both sensual and survival needs. The environment provides the economy with raw materials, transformed into consumer products through the production process and energy that fuels this transformation. These raw materials and energy return to the environment as waste (Terry and Lead, 2012). If the frame is set wide enough, the relationship between the environment and the economic system can be regarded as a closed system. A closed system is one in which the inputs are not taken from the outside and outputs cannot be transferred out of it—unlike an open system in which the materials or energy are imported or exported. Treating the planet as a closed system has one important consequence, which can be summarized in the first law of thermodynamics, in which energy and matter can be neither created nor destroyed. This law means that the material inputs in the economic system from the environment must be either accumulated into the financial system or returned to the environment as waste. If the accumulation of the mass of material inputs in the financial system stops, then the mass of waste flow to the environment must be equal. Any excessive amount of waste can reduce the value of environment as an asset. Therefore, when these quantities of waste exceed the abortive capacity of the environment, the value of the services that the environment provides to us as, is reduced.

Beyond that, the human-environment relationship is limited by another physical law, the second law of thermodynamics (Burness and Morris, 2014). Generally known as the law of entropy, this law states that the entropy increases. When applied in energy processes, this law means that if we transform one type of energy to another, this transformation will never be absolute and that energy consumption is an irreversible process. Some amount of energy is always lost during the conversion process while the remainder is lost in the final utilization. The second law of thermodynamics also explains that if there are no new inputs in a closed system, the available energy will eventually be exhausted.

On the other hand, the planet is not a closed system as the energy comes from the sun. However, the law of entropy implies that there is an upper limit in the flow of energy from the sun that can be harnessed. When the stocks of stored energy are exhausted, the amount of energy that can be used will be determined by the flow of energy and the amount that can be stored. So in the very long term, the process of economic growth will be limited by the availability of the solar energy and our ability to use it.

It is known that the Earth's climate has changed throughout history and these changes can be associated with minor variations of the Earth's orbit that affect the amount of solar energy the Earth receives. However, for centuries, the amount of atmospheric carbon dioxide had never been as high as today. The situation is rather serious, as the current global warming is mostly caused by human activity since the middle of the 20th century (Shaftel et al., 2019). Because of that, the United Nations Development Programme has created 17 Sustainable Development Goals, which refer to the action each nation should take in order to end poverty, ensure peace and prosperity, and protect the planet. Denmark has taken these Sustainable Development Goals and introduced them to its national energy and climate plan. Currently, it can be noticed that the Danish government is actively working to reach these targets. Based on the fact that CO_2 emissions are one of the main aspects that have an impact on global warming, it became crucial to determine whether Denmark has taken any actions on eliminating CO_2 emissions. Thus, the automotive industry in Denmark should be continuously analyzed. This will allow determining the current situation of the vehicle fleet in Denmark and the potential of hydrogen and alternative fuels in the market. Thereby, it will be possible to determine both current and a future number of vehicles that use or will use renewable energy. All in all, as the influence of renewable energy in the automotive industry in Denmark will be known, it will be possible to determine the effectiveness of achieving sustainable goals by Denmark (Roungkvist et al., 2020).

Over the past 20 years, the hydrogen economy was investigated with a focus on process optimization (Gibson and Kelly, 2008), storage improvement (Chen et al., 2009), automotive applications and transportation applications (Juste, 2016; Apostolou et al., 2021), green infrastructure (Apostolou et al., 2019) and economic aspects (Winter, 2009). During this period, more hydrogen applications were defined, creating a broad spectrum of production processes and finished product usage, depending on its application. Due to the continuously growing areas of H_2 applications, the research community needed a firm definition of the hydrogen economy. Over the past decade, the term Power to X (P2X) was introduced to ease the understanding of types of conversion's finished products and the production processes that come with it. P2X was defined as a notation of systems, which can convert the excess electric power into other forms of energy. The "X" in the term refers to the final product/form of energy of the above-mentioned conversion. This process can be based on different pathways where the end product could be gas, liquid, or heat (Lewandowska-Bernat and Desideri, 2017). On the other hand, P2X was also defined as a context describing all technological pathways from hydrogen production via electrolysis to the end user throughout the energy sector (Decourt, 2019). The main approaches of P2X to obtain both environmentally and economically beneficial utilization of renewable energy could be divided into demand-side management and electricity storage.

In 2010, Denmark announced the possibility of being fossil fuel independent. A long-term strategy, Energy Strategy 2050, was created, which demonstrates new energy policies for the reduction of fossil fuels and the increase of renewable energy. The initiatives presented by the government will eventually allow Denmark to have lower energy consumption, secure supply of energy, increase the share of renewable energy, and create new opportunities for Danish companies that operate in the climate and energy sector. This work focuses on drawing a parallel to the EV growth over the last decade and tries to make an educated guess on how the hydrogen growth will be in the next decade. Will the expected growth have an impact towards the greener Danish society?

10.2 Methodology

In order to determine whether hydrogen will influence the achievement of Denmark's sustainable goals, hydrogen and its properties should be well known. This research provided crucial information regarding hydrogen's properties and possible production methods. Thereafter, the advantages and disadvantages of hydrogen were defined, especially in its application in the automotive industry.

Thereafter, the current situation of fuel cell electric vehicles (FCEVs) in Denmark was analyzed. The number of cars fueled by hydrogen was found, as well as the number of hydrogen fueling stations (Apostolou and Xydis, 2019). This gave the information upon when has Denmark started to invest in hydrogen technology. The progress of the hydrogen technology implementation in Denmark could be determined (Rasmussen et al., 2020). Nevertheless, as hydrogen fuel cell vehicles are a relatively new technology in Europe, the assumption of its future development had to be made. Therefore, the electric vehicles (EV) have been analyzed in order to gather the data and determine their increase in sales over the years. Also, the increase in the number of charging stations was analyzed to determine whether there was any connection between the number of the stations and number of the EV registered. Once the relationship had been identified, it was assumed that this would also be the case for hydrogen cars. After the assumption, the historical data of the number of hydrogen cars and hydrogen fuel stations were found. This enabled us to predict the possible number of hydrogen cars by 2030. Thereby, it was also possible to calculate how much CO_2 could be eliminated only by hydrogen cars in 2030.

In the last part of this analysis, Denmark's action in connection to hydrogen promotion was discussed. The discussion reflected upon the number of hydrogen cars in the industry and in the private sector. Additionally, the discussion focused upon what Denmark could do in order to increase the awareness of hydrogen technology used in the automotive industry. The conclusion reflected upon the entire chapter and addressed the question of whether hydrogen will have an impact on the achievement of Denmark's sustainability targets.

10.3 Analysis

10.3.1 Hydrogen Properties and Application

In order to determine hydrogen's impact on the reduction of the carbon dioxide emission in Denmark, sources of hydrogen, its properties and production of hydrogen should have been acknowledged. Because of that, the literature overview upon hydrogen must be conducted. According to Mustafa Balat, "hydrogen is the lightest element, colourless, odourless, tasteless and nontoxic gas found in the air at concentrations of about 100 ppm (0.01%)" (Balat, 2008). Additionally, hydrogen is a gas at its standard temperature and pressure, which is 0°C and 1013 hPa. Nevertheless, more importantly, this element can be found everywhere as it represents 75 wt% (weight percent) and 90 vol% (volume percent) of all matter (Godula-Jopek, 2015). The lower heating value (LHV) of hydrogen has the energy content of 39.4 kWhkg^{-1}, which corresponds to 142 MJkg^{-1}. This means, that for each one kg of hydrogen, it has an energy value of 142 MJ (Godula-Jopek, 2015). The difference between the LHV and Upper Heating Value (called molar enthalpy of vaporization of water) is the amount of energy required to convert one mole of a substance from the liquid state to the gas state. This must be conducted at a constant both temperature and pressure (Helmenstine, 2017). Additionally, hydrogen has an extremely small atomic weight. Because hydrogen has very low density—e.g., hydrogen's liquid density is 10 times smaller than gasoline's—this indicates that hydrogen is able to store more energy per unit in comparison to the gasoline but not more volume. Storing hydrogen is the main issue for which hydrogen application as a transportation fuel is not as advanced as it should or could have been. Hydrogen can be stored in a vehicle as compressed gas, a liquid in a cryogenic container or as gas bound with specific metals. Nevertheless, due to the low density, compressed hydrogen would require a large storage tank; otherwise, it becomes problematic to ensure a large driving range similar to gasoline. However, under high pressures, it has been a commercial fuel for transportation and a number of car industries are introducing fuel cell models based on hydrogen. On the other hand, hydrogen as gas bound with other metals and metal hydrides increases the weight of the energy source. This makes it difficult while applying it in the vehicle (Balat, 2008). On the contrary, hydrogen as a transportation fuel can be produced from numerous renewable resources, which are accessible worldwide. Additionally, it can be used as direct fuel in a vehicle, which does not differ largely from the engines that use gasoline. The engines that use hydrogen as a transportation fuel are said to start easily at lower temperatures, as hydrogen can last in its gas form until –253.15°C. Most importantly, hydrogen is one of the transportation fuels that do not emit CO_2 (Balat, 2008).

However, hydrogen itself it is not an energy source and in that manner, it must be obtained by conversion processes of different sources. These sources can be, e.g., fossil fuels and renewable sources or nuclear power (Godula-Jopek, 2015). There are four categories of hydrogen-extracting process from natural resources: (1) thermal, (2) electrical, (3) photonic and (4) biochemical (Levin et al., 2004). However, all methods of generating energy from green energy sources processes are executed in different environments, especially in different temperatures. Thermochemical, thermocatalytic, thermolysis, and high-temperature electrolysis demand high-temperature heat. Additionally, solar energy and biomass energy are also able to generate high-temperature heat. However, hybrid thermochemical water splitting and hydrogen sulfide splitting require intermediate temperature. Heat recovered from various processes as well as solar, geothermal and nuclear are processes that also require intermediate heat. Thermophilic is a process that demands low heat temperature. Another aspect that must be considered in these processes is the fact, that some of them require additional input. An example of these processes is photonic energy, which requires the input of solar energy (Xydis, 2013). Another example are the biological conversion processes that require biomass as a substrate (Dincer, 2011). In vehicles, compressed hydrogen gas is used to generate electric power through the energy converter, which is the fuel cell. The fuel cell then transforms hydrogen into electricity, used thereafter to power the engine. The vehicles that use hydrogen to produce power for the engine are called fuel cell electric vehicles (FCEV) (Davies, 2013). Additionally, in FCEVs, cars can be built either as a series hybrid or as a parallel hybrid. A series hybrid requires a small fuel cell but a large battery that drives the motor all the time. The battery is continuously charged by the fuel cells, whereas the parallel hybrid has a large fuel cell tank and a rather small battery, which adds power when needed.

10.3.2 Hydrogen Vehicle's Current Situation in Denmark

Based on the research, the first hydrogen refueling station was built in 2011 in Holstebro. This station was a 700 bar station that enabled to fuel a car in less than 3 min, which is extremely convenient as the time used is similar to fueling petrol car. However, in 2016 Denmark became the first country with a national network of hydrogen fueling stations. Until that year, there were six H_2 stations with on–site electrolyzer, three H_2 stations where hydrogen is delivered from central electrolyzer plant operated by Strandmøllen, one planned and one H_2 station under consumption. All these stations were compatible with the current technology of hydrogen fuel cell cars, as the pressure of each station equaled 70 MPa (Kane, 2016). According to Ludwig-Bölkow-Systemtechnik GmbH, there are currently nine operational hydrogen stations in Denmark (and four older projects) and some more planned to be built in the following years (H_2 stations, 2021). The map showing the placement of both operational and planned stations is presented in Fig. 10.1.

The number of these stations will increase in the following years, as the company NEL Hydrogen, which installs hydrogen stations in Denmark, has stated that their aim is to "ensure that 50% of Danish population will have less than 15 km to hydrogen fueling". (H_2 Logic delivers 9th H_2 fueling station for Denmark, 2016). Several hydrogen vehicle models are available on the Danish market since 2013. According to the website dedicated

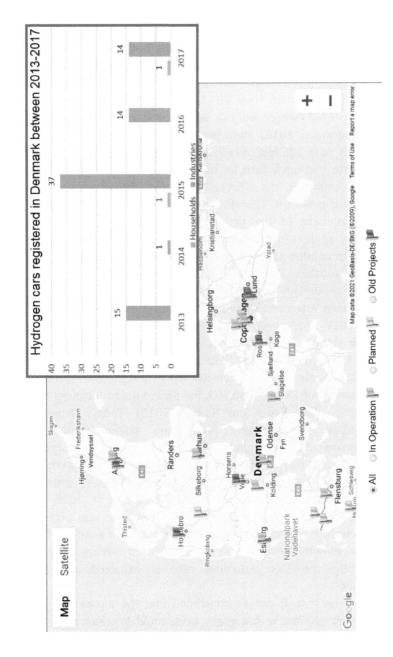

Figure 10.1 Operational and planned hydrogen stations.

to hydrogen cars in Denmark, https://brintbiler.dk, there are three hydrogen car models available from two manufacturers. These are Hyundai iX35, Hyundai NEXO and Toyota Mirai (Biler, 2020). It seems that Honda Clarity is not available in Denmark anymore; it used to be available since 2018. It has discovered though that hydrogen passenger cars in Denmark were registered by the industry section since 2013. Until 2017, 83 hydrogen cars were registered, where only 2 were registered in the private sector (Ingemann, 2018). Until the end of 2019, this number has grown only up to 105 fuel cell EVs (EAFO, 2021).

Because hydrogen vehicles are a rather new technology for European countries, the citizens might, in a broad perspective, not be familiar with it. Therefore, it became crucial to investigate the development of the previous innovative and sustainable technology applied in the automotive industry in Denmark since linking it to another technology with prospects, such as hydrogen, seems reasonable. The latest innovative technology preceding hydrogen technology is the electric vehicles (EV). According to the European Alternative Fuels Observatory, the battery electric vehicle (BEV) passenger cars started to being purchased in Denmark already in 2009. Continuing, in 2012 there were in a total of 453 normal and fast public charging points in Denmark (Alternative Fuels, 2019), which has led to a significant increase in the number of EVs in Denmark for personal and business use. (Electricity, 2020; Xydis and Nanaki, 2015). Today, Denmark has reached 350 charging points per million population (approximately 2,000 charging points, including new fast chargers) (Haustein et al., 2021). It should also be stressed that only a couple years ago the forecast was not expecting such a steep growth; however, based on the number of registered cars in 2020, the expected growth curve became steeper. If we take into account that Denmark is expecting that by 2030 it will have a million EVs, then the growth—several years after the infrastructure grew significantly—the EVs unfolding rate is expected to grow vigorously.

Based on that, it can be concluded that the increase of the Danish fleet of electric passenger cars, could be caused by the increase in the number of charging stations. Therefore, looking at the increase in the number of EV in Denmark, it can be noticed that since 2011 the number of EV in Denmark has more than doubled. Additionally, since one of the first charging stations has been installed in Denmark, the number of EV charging stations has increased more than five times. The data are presented in Fig 10.2. Lastly, based on the gathered information, it can be assumed that the numbers will continue to increase in the upcoming years.

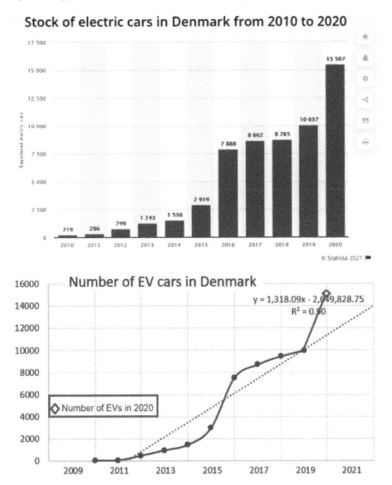

Figure 10.2 Number of EV cars in Denmark (DK).

Based on this assumption, the potential increase in both the number of hydrogen charging stations and hydrogen cars can be estimated. Based on the data gathered from Statistics Denmark, it can be concluded that the number of hydrogen passenger car is fluctuating. Although the first fueling station was built in 2011, there were no hydrogen cars registered until 2013. However, 2013 was the year when the first hydrogen cars became available in the Danish market. Since that year, the number of hydrogen passenger cars has increased and decreased, which indicates that the number of cars has been fluctuating. However, since 2016 the number of these cars has become more stable, which can indicate that from that year the sales rate of the cars will be steadily increasing. If that will be the case, then in 2030 there will be approximately 4,000 hydrogen cars in total (Fig. 10.3).

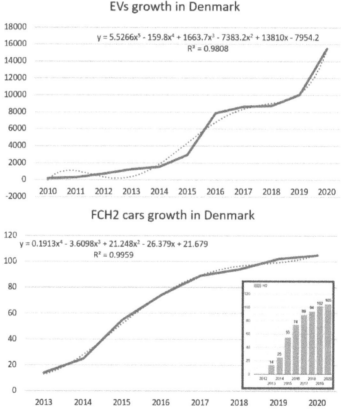

Figure 10.3 Projection for the total number of EV and H_2 cars in Denmark (DK).

What should be stressed though is that it took at least 10–15 years from setting up the EV charging infrastructure to establish significant growth in 2020. However, due to political will, it is not expected to take as long as in the case of EVs. It seems that there is the will for establishing a more aggressive growth strategy for hydrogen cars. This can be supported by the fact that in Denmark large hydrogen companies have their headquarters, such as NEL hydrogen (EnerFuel), Blue World Technologies, Ballard, etc., and these companies have committed their progress on hydrogen innovation and future investments. The risks for hydrogen growth are also lowered further due to the fact that hydrogen is a cheaper fuel than gasoline on absolute numbers. However, since this cost is associated to the cost of hydrogen vehicles, this cannot be immediately seen. Another parameter is that transportation has been forever a gas-based running economy on the contrary to the electric vehicle business model. This means that a hydrogen-based model will be based not only on the electricity sector but on the cleaner gas-based technology, such as hydrogen. Last, on a larger scale, hydrogen can be the perfect clean fuel for larger transportations (train, bus, ferry, airplane), where EVs are not ready and perhaps will never be (Nanaki et al., 2015). Having said all this, it is clear that the development of a hydrogen-reliant transportation sector involved fewer risks.

10.4 Results and Discussion

The larger the number of cars that use renewable energy, the greater the elimination of CO_2 emission. Assuming that an average petrol or diesel car emits 118.5 g CO_2/km and assuming that each car drives 10,000 km yearly, each car that uses renewable energy is able to eliminate 1,185,000 g CO_2 yearly. However, based on the data from Statistics Denmark, from 2011 until 2020, a total of 15,507 new renewable cars were registered. This means that by 2020 Denmark was able to eliminate 18,375,795 kg CO_2 annually. This will not be enough to meet the sustainability standards, as the number of cars that use renewable energy is less than 1% of the total number of new cars registered during that time (Ingemann, 2018). Additionally, from a "hydrogen" point of view, by the year 2030, assuming that the increase in new registration of passenger cars will be according to the prediction (Fig. 10.3), hydrogen cars alone will be able to eliminate 4,740,000 kg CO_2 annually. Clearly, the fact that there is an increase in the number of renewable cars is excellent, yet the scope of the increase might not be satisfactory in order to meet sustainable targets.

As described, Denmark has first taken the initiative in creating a reliable environment for the possession of the hydrogen car, and thereafter it has set the example by purchasing them. Nevertheless, in order to fulfill all the targets, all sectors must be involved in expanding the usage of technology. This is currently not the case, as only a couple of hydrogen cars have been purchased by the private sector. This indicates that in order to meet the country's sustainable targets, there should be more private buyers. This can be achieved only if the citizens become more aware of the availability of the technology and the required infrastructure is offered—in other words, the chicken-and-egg problem again, exactly as in the EV growth case (Apostolou and Welcher, 2021). On top of that, in terms of technology, and because there is a common misconception that hydrogen is dangerous, the awareness of the advantages of hydrogen cars, as well as the safety measures and precautions that were taken into consideration while development, should be explained and advertised, in order to familiarize people with the technology.

10.5 Conclusion

This research supported the hypothesis that the Danish energy and transportation sector will become greener by massively introducing hydrogen by following a more aggressive growth strategy for hydrogen cars. This is stated based on the investment and declarations the Danish government has taken in order to ensure that all sustainable targets will be met. However, despite the fact that the use of hydrogen technologies is increasing in Denmark, the rate of increase is not high enough. Therefore, it can be stated that hydrogen technologies can influence Denmark's achievement of its sustainability goals only if the number of hydrogen cars increases. The larger the number of hydrogen cars, the smaller amount of CO_2 emissions. Combining the number of cars that use renewable energy in Denmark revealed that they are currently only 1% of the total number of cars in Denmark. To conclude, the number of cars that use renewable energy must increase in the coming years if Denmark wants to meet its strategic targets. Increasing the investment based on political will not be as risky as investing in the EV infrastructure once was. The framework for green growth is already in place and the hydrogen development speed can be higher, minimizing the associated risks. However, this can only be achieved if Danish citizens are more informed about technology. If they are not properly informed, they may react negatively because they will not feel confident, nor safe. This could prevent Denmark from meeting its sustainability targets.

References

1. Alternative Fuels, Electricity. Retrieved 12 05, 2018, from European Alternative Fuels Observatory: https://www.eafo.eu/alternative-%20 fuels/electricity/charging-infra-stats.

2. Apostolou, D., Casero, P., Gil-Hernández, V., Xydis, G., Integration of a light mobility urban scale hydrogen refuelling station for cycling purposes in the transportation market, *International Journal of Hydrogen Energy*, 47(7), 2021, pp. 5756–5762, DOI: 10.1016/ j.ijhydene.2020.11.047.

3. Apostolou, D., Enevoldsen, P., Xydis, G., Supporting green urban mobility: The case of a small-scale autonomous hydrogen refuelling station, *International Journal of Hydrogen Energy*, 44(20), 2019, pp. 9675–9689, Special issue of SI:ICH$_2$P-2018, DOI: 10.1016/ j.ijhydene.2018.11.197.

4. Apostolou, D., Xydis, G., A literature review on hydrogen refuelling stations and infrastructure. *Current Status and Future Prospects, Renewable & Sustainable Energy Reviews*, 113, 2019, p. 109292, DOI: 10.1016/j.rser.2019.109292.

5. Apostolou, D., Welcher, S., Prospects of the hydrogen-based mobility in the private vehicle market. A social perspective in Denmark, *International Journal of Hydrogen Energy*, 46(9), 2021, pp. 6885–6900.

6. Balat, M., Potential importance of hydrogen as a future solution to environmental and transportation problems, *International Journal of Hydrogen Energy*, 33(15), 2008, pp. 4013–4029.

7. Biler, 2020, Retrieved from BrintBiler: https://brintbiler.dk.

8. Burness, S., Morris, G., Thermodynamic and concepts as related to resource use policies, Land Economics, 2014.

9. Chen, H., Cong, T. N., Yang, W., Tan, C., Li, Y., Ding, Y., Progress in electrical energy storage system: A critical review, *Progress in Natural Science*, 19(3), 2009, pp. 291–312.

10. Davies, A. A hydrogen-powered car could someday be sitting in your driveway, 2013. Retrieved 12 04, 2018, from Business Insider: https://www.businessinsider.com/pro-con-future-of-hydrogen-fuel-cell-electric-vehicles-fcev-2013-4?r=US&IR=T.

11. Decourt, B., Weaknesses and drivers for power-to-X diffusion in Europe. Insights from technological innovation system analysis, *International Journal of Hydrogen Energy*, 44(33), 2019, pp. 17411–17430.

12. Dincer, I., Green methods for hydrogen production. *International Journal of Hydrogen Energy*, 37, 2011, pp. 1954–1971.

13. EAFO. Passenger cars - Number of alternative fuels passenger cars-H_2 2021. https://www.eafo.eu/vehicles-and-fleet/m1#.

14. Gibson, T. L., Kelly, N. A., Optimization of solar powered hydrogen production using photovoltaic electrolysis devices, *International Journal of Hydrogen Energy*, 33(21), 2008, pp. 5931–5940.

15. Godula-Jopek, A., Chapter 1—Introduction. In Godula-Jopek, A., Stolten, D., *Hydrogen Production: By Electrolysis*, John Wiley & Sons, Incorporated, 2015, pp. 1–32.

16. Haustein, S., Jensen, A. F., Cherchi, E., Battery electric vehicle adoption in Denmark and Sweden: Recent changes, related factors and policy implications, *Energy Policy*, 149, 2021, p. 112096.

17. H_2 stations, H_2 stations map, 2021, Retrieved: https://www.H_2 stations.org/stations-map/?lat=49.763948&lng=12.582221&zoom=4.

18. H_2 Logic delivers 9th H_2 fueling station for Denmark; 100% renewable hydrogen, 1st country-wide station network, 2016, Retrieved: Dec. 2020, Available at: https://www.greencarcongress.com/2016/03/20160304-H_2logic.html.

19. Helmenstine, A. M. Molar enthalpy of vaporization definition, (2017, 03 06). Retrieved 12.03, 2018, from ThoughtCo.: https://www.thoughtco.com/definition-of-molar-enthalpy-of-vaporization-605361.

20. Ingemann, K. M. BIL51: New registrations of passenger cars by ownership and propellant, 2018. Retrieved 12 05, 2018, from Statistics Denmark: https://www.statbank.dk/BIL51.

21. Juste, G. L., Hydrogen injection as additional fuel in gas turbine combustor. Evaluation of effects, *International Journal of Hydrogen Energy*, 31(14), 2006, pp. 2112–2121.

22. Kane, M. With 9 hydrogen fuel stations, Denmark is 1st country with basic National Network, 2016. Retrieved 12 04, 2018, from https://insideevs.com/news/328879/with-9-hydrogen-fuel-stations-denmark-is-1st-country-with-basic-national-network.

23. Levin, D. B., Pitt, L., Love, M., Biohydrogen production: Prospects and limitations to practical application. *International Journal of Hydrogen Energy*, 29(2), 2004, pp. 173–185.

24. Lewandowska-Bernat A., Desideri, U., Opportunities of power-to-gas technology in different energy systems architectures, *Applied Energy*, 228, 2017, pp. 57–67.

25. Nanaki, E. A., Xydis, G. A., Koroneos, C. J., Electric vehicle deployment in Urban areas, *Indoor and Built Environment* (special issue: Urban Sustainability), Sage Publications, 25(7), 2015, pp. 1065–1074, DOI: 10.1177/1420326X15623078.

26. Shaftel, H., Jackson, R., Callery, S., Climate change: How do we know? (2019, 12 03). Retrieved 12 06, 2018, from global climate change vital signs of the planet: https://climate.nasa.gov/evidence/.

27. Terry, L., Leal, D., Free market environmentalism, Westview Press, 2012.

28. Trout, C. Denmark lands Europe's first better place EV battery swapping station, (2011, 07 01). Retrieved 2020: https://www.engadget.com/2011-07-01-denmark-lands-europes-first-better-place-ev-battery-swapping-st.html.

29. Rasmussen, N. B., Enevoldsen, P., Xydis, G., Transformative multi-value business models: A bottom-up perspective on the hydrogen-based green transition for modern wind power cooperatives, *International Journal of Energy Research*, 44(5), 2020, pp. 3990–4007, DOI: 10.1002/ER.5215.

30. Roungkvist, J. S., Enevoldsen, P., Xydis, G., High-resolution electricity spot price forecast for the Danish power market, *Sustainability, SI: Sustainable and Renewable Energy Systems, Sustainability*, 12, 2020, p. 4267. DOI: 10.3390/su12104267.

31. Winter, C. J., Hydrogen energy–Abundant, efficient, clean: A debate over the energy-system-of-change, *International Journal of Hydrogen Energy*, 34(14) SUPPL. 1, 2009, pp. 1–52.

32. Xydis, G., The wind chill temperature effect on a large-scale PV plant: An exergy approach, *Progress in Photovoltaics: Research and Applications*, DOI: 10.1002/pip.2247.

33. Xydis, G., Nanaki, E., Wind energy based electric vehicle charging stations sitting, *A GIS/Wind Resource Assessment Approach Challenges*, 6(2), 2015, pp. 258–270, DOI: 10.3390/challe6020258.

Index

For Product Safety Concerns and Information please contact our EU
representative GPSR@taylorandfrancis.com
Taylor & Francis Verlag GmbH, Kaufingerstraße 24, 80331 München, Germany